AN INTRODUCTION TO AGRICULTURAL ENGINEERING

Edited by
Barbara June Dinampo

ArclerPress

An Introduction to Agricultural Engineering

Edited by Barbara June Dinampo

Edition 2018

ArclerPress

Published by
Arcler Press LLC
708 3rd Avenue, 6th Floor
New York
NY 10017
United States of America
www.arclerpress.com

ISBN: 978-1-68094-542-3 (Hardback)

Printed in United States of America

Library of Congress Control Number: 2017933114

About the Editor

Barbara June Dinampo

Barbara June is an Agricultural and Biosystems Engineer. She obtained her degree from Mindanao State University-Marawi, Philippines in 2007. She has been working on Agricultural project and development focused on food security in a government agency. Currently, she is a freelance writer and editor of Agricultural journal, textbooks, and references.

Table of Contents

2 WATER USE AND LAND USE

3 IRRIGATION MANAGEMENT

4 SURVEYING

5 SEED MANAGEMENT PRACTICES

6 TILLAGE SYSTEMS AND TECHNIQUES

7 LIVESTOCK MANAGEMENT PRACTICES

8 ENVIRONMENTAL IMPACT OF AGRICULTURE

List of Abbreviations

ACSM	American Congress on Surveying and Mapping
ADB	Asian Development Bank
BLM	Bureau of Land Management
EDM	Electronic Distance Measurement
FAO	Food and Agriculture Organization
FT	Forest Transition
GLO	General Land Office
GIS	Geographical Information Systems
GPS	Global Positioning System,
LFHD	Low Forest Cover with High Rates of Deforestation
NPS	Nonpoint Source
PLSS	Public Land Survey System
RTK	Real Time Kinematic
REDD	Reducing Emissions from Deforestation and Forest Degradation

Preface

Agricultural Engineering is the area of engineering concerned with the design, construction and improvement of farming equipment and machinery. Agricultural engineers may perform tasks such as planning, supervising and managing the building of dairy effluent schemes, irrigation, drainage, flood water control systems, performing environmental impact assessments, agricultural product processing and interpret research results and implement relevant practices. Agricultural engineers integrate technology with farming. For example, they design new and improved farming equipment that may work more efficiently, or perform new tasks. They design and build agricultural infrastructure such as dams, water reservoirs, warehouses, and other structures. They may also help engineer solutions for pollution control at large farms. Some agricultural engineers are developing new forms of biofuels from non-food resources like algae and agricultural waste. Such fuels could economically and sustainably replace gasoline without jeopardizing the food supply. For decades, agriculture has been associated with production of essential food crops. At present, agriculture above and beyond farming includes forestry, dairy, fruit cultivation, poultry, bee keeping, mushroom, arbitrary, etc. Today, processing, marketing and distribution of crops and livestock products etc. are all acknowledged as part of current agriculture. Thus, agriculture could be referred to as the production, processing, promotion and distribution agricultural products. Agriculture plays a critical role in the entire life of a given economy. Agriculture is the backbone of economic system of a given country. In addition to providing food and raw material, agriculture also provides employment opportunities to very large percentage of population. Agricultural engineers must have a wealth of knowledge and skills to function effectively in the multiplicity of the agricultural and agribusiness industries. The agricultural engineer helps to make farming sustainable, safe, and environmentally friendly. They analyze agricultural operations and weigh the use of new technologies and methods to increase yields, improve land use, and conserve resources like seed, water, fertilizers, pesticides and fuel. The engineer recommends strategies to protect the health, safety and security of worker's, animals and agricultural products.

Agricultural engineers develop methods and design equipment for land preparation, planting and harvesting. They incorporate automation, precision and smart technologies "intelligence" to new and existing equipment. Sensors (including biosensors) are combined with microcomputers, controllers, artificial intelligence and other software, to optimize efficiency, sustainability, and the reliability of our food, feed, fiber and fuel economy. Agriculture engineers find better ways to reduce crop loss from field damage, during handling, sorting, packing and processing. The warehousing of food and fiber are an important part of the agriculture industry. The agricultural engineer is the person who plans the heating, cooling, ventilation, postharvest handling, logistics and more.

Content Coverage

Chapter One introduces about agricultural machinery used in farming or other agriculture. There are many types of such equipment, from hand tools and power tools to tractors and the countless kinds of farm implements that they tow or operate. Diverse arrays of equipment are used in both organic and nonorganic farming.

Chapter Two focuses on water and land use concepts for the production of a given quantity of some product or crop.

Chapter Three focuses on irrigation techniques used to assist in the growing of agricultural crops, maintenance of landscapes, and revegetation of disturbed soils in dry areas and during periods of inadequate rainfall. Additionally, irrigation also has a few other uses in crop production, which include protecting plants against frost, suppressing weed growth in grain fields and preventing soil consolidation.

Chapter Four presents the basics of land surveying; the technique, profession, and science of determining the terrestrial or three-dimensional position of points and the distances and angles between them.

Chapters Five and *Six* emphasize the seed management practices and tillage systems, respectively. A combination of technique, equipment, pesticides, crop rotation, fertilization, and irrigation have to be used for local conditions.

Chapter Seven covers practices of livestock management as animal husbandry practices have varied widely across cultures and time periods. Originally, livestock were not confined by fences or enclosures, but these practices have largely shifted to intensive animal farming, sometimes referred to as "factory farming".

Chapter Eight highlights about the environmental impact of agriculture that involves a variety of factors from the soil, to water, the air, animal and soil diversity, people, plants, and the food itself. Some of the environmental issues that are related to agriculture are climate change, deforestation, genetic engineering, irrigation problems, pollutants, soil degradation, and waste.

CHAPTER 1

Agricultural Machinery and Equipment

Chapter Contents

- **Introduction**
- **History of the Machines**
- **Types**
- **New Technology and the Future**
- **Combine Harvester**
- **Cultivator**
- **Baler**
- **Beaverslide**

Agricultural technology is the tools and machinery that are used primarily or entirely in order to support agricultural enterprise. Examples include plows, threshers, and irrigation systems. These forms of technology, which are all regularly used in modern agricultural settings, have a long history in farming and have been reinvented and redesigned many times over. Plows, for example, were originally pulled by animals but now are usually powered by engines. For people who are interested in the history of agricultural technology, there are a number of museums that have preserved old or outdated pieces of machinery and have staff on hand to explain how they worked and why they have been replaced.

INTRODUCTION

Agricultural machinery is machinery used in farming or other agriculture. There are many types of such equipment, from hand tools and power tools to tractors and the countless kinds of farm implements that they tow or operate. Diverse arrays of equipment are used in both organic and nonorganic farming. Especially since the advent of mechanised agriculture, agricultural machinery is an indispensable part of how the world is fed.

A German combine harvester by Claas

A British crop sprayer **by** Lite-Trac.

HISTORY OF THE MACHINES

The Industrial Revolution

With the coming of the Industrial Revolution and the development of more complicated machines, farming methods took a great leap forward.Instead of harvesting grain by hand with a sharp blade, wheeled machines cut a continuous swath. Instead of threshing the grain by beating it with sticks, threshing machines separated the seeds from the heads and stalks. The first tractors appeared in the late 19th century.

Steam Power

Power for agricultural machinery was originally supplied by ox or other domesticated animals. With the invention of steam power came the portable engine, and later the traction engine, a multipurpose, mobile energy source that was the ground-crawling cousin to the steam locomotive. Agricultural steam engines took over the heavy pulling work of oxen, and were also equipped with a pulley that could power stationary machines via the use of a long belt. The steam-powered machines were low-powered by today`s standards but, because of their size and their low gear ratios, they could provide a large drawbar pull. Their slow speed led farmers to comment that tractors had two speeds: "slow, and damn slow."

Internal Combustion Engines

The internal combustion engine; first the petrol engine, and later diesel engines; became the main source of power for the next generation of tractors. These engines also contributed to the development of the self-propelled, combined harvester and thresher, or combine harvester (also shortened to 'combine'). Instead of cutting the grain stalks and transporting them to a stationary threshing machine, these combines cut, threshed, and separated the grain while moving continuously through the field.

TYPES

A John Deere cotton harvester at work in a cotton field.

From left to right: John Deere 7800 tractor with Houle slurry trailer, Case IH combine harvester, New Holland FX 25 forage harvester with corn head.

A New Holland TR85 combine harvester.

Combines might have taken the harvesting job away from tractors, but tractors still do the majority of work on a modern farm. They are used to push implements—machines that till the ground, plant seed, and perform other tasks.

Tillage implements prepare the soil for planting by loosening the soil and killing weeds or competing plants. The best-known is the plow, the ancient implement that was upgraded in 1838 by John Deere. Plows are now used less frequently in the U.S. than formerly, with offset disks used instead to turn over the soil, and chisels used to gain the depth needed to retain moisture.

The most common type of seeder is called a planter, and spaces seeds out equally in long rows, which are usually two to three feet apart. Some crops are planted by drills, which put out much more seed in rows less than a foot apart, blanketing the field with crops. Transplanters automate the task of transplanting seedlings to the field. With the widespread use of plastic mulch, plastic mulch layers, transplanters, and seeders lay down long rows of plastic, and plant through them automatically.

After planting, other implements can be used to cultivate weeds from between rows, or to spread fertilizer and pesticides. Hay balers can be used to tightly package grass or alfalfa into a storable form for the winter months.

Modern irrigation relies on machinery. Engines, pumps and other specialized gear provide water quickly and in high volumes to large areas of land. Similar types of equipment can be used to deliver fertilizers and pesticides.

Besides the tractor, other vehicles have been adapted for use in farming, including trucks, airplanes, and helicopters, such as for transporting crops and making equipment mobile, to aerial spraying and livestockherd management.

NEW TECHNOLOGY AND THE FUTURE

The basic technology of agricultural machines has changed little in the last century. Though modern harvesters and planters may do a better job or be slightly tweaked from their predecessors, the US$250,000 combine of today still cuts, threshes, and separates grain in the same way it has always been done. However, technology is changing the way that humans operate the machines, as computer monitoring systems, GPS locators, and self-steer programs allow the most advanced tractors and implements to be more precise and less wasteful in the use of fuel, seed, or fertilizer. In the foreseeable future, there may be mass production of driverless tractors, which use GPS maps and electronic sensors.

Open Source Agricultural Equipment

Many farmers are upset by their inability to fix the new types of high-tech farm equipment. This is due mostly to companies using intellectual property law to prevent farmers from having the legal right to fix their equipment (or gain access to the information to allow them to do it). This has encouraged groups such as Open Source Ecology and Farm Hack to begin to make open source agricultural machinery. In addition on a smaller scale FarmBot and the RepRap open source 3D printer community has begun to make open-source farm tools available of increasing levels of sophistication. In October 2015 an exemption was added to the DMCA to allow inspection and modification of the software in cars and other vehicles including agricultural machinery.

COMBINE HARVESTER

A Lely open-cab combine.

Harvesting oats with a Claas Lexion 570 harvester with enclosed air-conditioned cab, rotary thresher and laser-guided automatic steering.

Old Style Harvester found in the Henty, Australia region

John Deere Combine 9870 STS with 625D

Rostselmash Combine Torum 740

John Deere 9870 STS underbelly

Case IH Axial-Flow combine.

The **combine harvester**, or simply **combine**, is a machine that harvests grain crops. The name derives from its combining three separate operations comprising harvesting—reaping, threshing, and winnowing—into a single process. Among the crops harvested with a combine are wheat, oats, rye, barley, corn (maize), sorghum, soybeans, flax (linseed), sunflowers, and canola. The waste straw left behind on the field is the remaining dried stems and leaves of the crop with limited nutrients which is either chopped and spread on the field or baled for feed and bedding for livestock.

Combine harvesters are one of the most economically important labour saving inventions, significantly reducing the fraction of the population that must be engaged in agriculture.

Scottish inventor Patrick Bell invented the reaper in 1826. The combine was invented in the United States by Hiram Moore in 1834. Early versions were pulled by horse, mule or ox teams. In 1835, Moore built a full-scale version and by 1839, over 20 ha (50 acres) of crops were harvested. By 1860, combine harvesters with a cutting, or swathe, width of several metres were used on American farms. Australian Hugh Victor McKay produced a commercially successful combine harvester in 1885, the Sunshine Harvester.

Combines, some of them quite large, were drawn by mule or horse teams and used a bullwheel to provide power. Later, steam power was used, and George Stockton Berry integrated the combine with a steam engine using straw to heat the boiler. At the turn of the twentieth century, horse drawn combines were starting to be used on the American plains and Idaho (often pulled by teams of twenty or more horses).

In 1911, the Holt Manufacturing Company of California produced a self-propelled harvester. In Australia in 1923, the patented Sunshine Auto Header was one of the first center-feeding self-propelled harvesters.In 1923 in Kansas, the Baldwin brothers and their Gleaner Manufacturing Company patented a self-propelled harvester that included several other modern improvements in grain handling. Both the Gleaner and the Sunshine used Fordson engines; early Gleaners used the entire Fordson chassis and driveline as a platform. In 1929 Alfredo Rotania of Argentina patented a self-propelled harvester. International Harvester started making horse-pulled combines in 1915. At the time horse powered binders and stand alone threshing machines were more common. In the 1920s Case Corporation and John Deere made combines and these were starting to be tractor pulled with a second engine aboard the combine to power its workings. The world economic collapse in the 1930s stopped farm equipment purchases thus people largely retained the older method of harvesting. A few farms did invest and used Caterpillar tractors to move the outfits.

Tractor-drawn combines (also called pull-type combines) became common after World War II as many farms began to use tractors. An example was the All-Crop Harvester series. These combines used a shaker to separate the grain from the chaff and straw-walkers (grates with small teeth on an eccentric shaft) to eject the straw while retaining the grain. Early tractor-drawn combines were usually powered by a separate gasoline engine, while later models were PTO-powered. These machines either put the harvested crop into bags that were then loaded onto a wagon or truck, or had a small bin that stored the grain until it was transferred to a truck or wagon with an auger.

In the U.S., Allis-Chalmers, Massey-Harris, International Harvester, Gleaner Manufacturing Company, John Deere, and Minneapolis Moline are past or present major combine producers. In 1937, the Australian-born Thomas Carroll, working for Massey-Harris in Canada, perfected a self-propelled model and in 1940 a lighter-weight model began to be marketed widely by the company. Lyle Yost invented an auger that would lift grain out of a combine in 1947, making unloading grain much easier. In 1952 Claeys launched the first self-propelled combine harvester in Europe; in 1953, the European manufacturer Claas developed a self-propelled combine harvester named 'Hercules', it could harvest up to 5 tons of wheat a day. This newer kind of combine is still in use and is powered by diesel or gasoline engines. Until the self-cleaning rotary screen was invented in the mid-1960s combine engines suffered from overheating as the chaff spewed out when harvesting small grains would clog radiators, blocking the airflow needed for cooling.

A significant advance in the design of combines was the rotary design. The grain is initially stripped from the stalk by passing along a helical rotor instead of passing between rasp bars on the outside of a cylinder and a concave. Rotary combines were first introduced by Sperry-New Holland in 1975.

In about the 1980s on-board electronics were introduced to measure threshing efficiency. This new instrumentation allowed operators to get better grain yields by optimizing ground speed and other operating parameters.

A New Holland TX68 with grain platform attached.

A John Deere Titan series combine unloading corn.

Combine Heads

Combines are equipped with removable heads that are designed for particular crops. The standard header, sometimes called a grain platform, is equipped with a *reciprocating knife cutter bar*, and features a revolving reel with metal or plastic teeth to cause the cut crop to fall into the auger once it is cut. A variation of the platform, a "flex" platform, is similar but has a cutter bar that can flex over contours and ridges to cut soybeans that have pods close to the ground. A flex head can cut soybeans as well as cereal crops, while a rigid platform is generally used only in cereal grains.

Some wheat headers, called "draper" headers, use a fabric or rubber apron instead of a cross auger. Draper headers allow faster feeding than cross augers, leading to higher throughputs due to lower power requirements. On many farms, platform headers are used to cut wheat, instead of separate wheat headers, so as to reduce overall costs.

Dummy heads or pick-up headers feature spring-tined pickups, usually attached to a heavy rubber belt. They are used for crops that have already been cut and placed in windrows or swaths. This is particularly useful in northern climates such as western Canada where swathing kills weeds resulting in a faster dry down.

While a grain platform can be used for corn, a specialized corn head is ordinarily used instead. The corn head is equipped with snap rolls that strip the stalk and leaf away from the ear, so that only the ear (and husk) enter the throat. This improves efficiency dramatically since so much less material must go through the cylinder. The corn head can be recognized by the presence of points between each row.

Occasionally rowcrop heads are seen that function like a grain platform, but have points between rows like a corn head. These are used to reduce the amount of weed seed picked up when harvesting small grains.

Self-propelled Gleaner combines could be fitted with special tracks instead of tires or tires with tread measuring almost 10in deep to assist in harvesting rice. Some combines, particularly pull type, have tires with a diamond tread which prevents sinking in mud. These tracks can fit other combines by having adapter plates made.

Conventional Combine

The cut crop is carried up the feeder throat (commonly called the "feederhouse") by a *chain and flight elevator*, then fed into the threshing mechanism of the combine, consisting of a rotating *threshing drum* (commonly called the "cylinder"), to which grooved steel bars (rasp bars) are bolted. The rasp bars thresh or separate the grains and chaff from the straw through the action of the cylinder against the *concave*, a shaped "half drum", also fitted with steel bars and a meshed grill, through which grain, chaff and smaller debris may fall, whereas the straw, being too long, is carried through onto the *straw walkers*. This action is also allowed because grain is heavier than straw, which causes it to fall rather than "float" across from the cylinder/concave to the walkers. The drum speed is variably adjustable on most machines, whilst the distance between the drum and concave is finely adjustable fore, aft and together, to achieve optimum separation and output. Manually engaged *disawning plates* are usually fitted to the concave. These provide extra friction to remove the awns from barley crops. After the primary separation at the cylinder, the clean grain falls through the concave and to the shoe, which contains the chaffer and sieves. The shoe is common to both conventional combines and rotary combines.

Hillside Leveling

Palouse hills

A Massey Ferguson combine fitted with the hillside leveling option.

In the Palouse region of the Pacific Northwest of the United States the combine is retrofitted with a hydraulic hillside leveling system. This allows the combine to harvest the steep but fertile soil in the region. Hillsides can be as steep as a 50% slope. Gleaner, IH and Case IH, John Deere, and others all have made combines with this hillside leveling system, and local machine shops have fabricated them as an aftermarket add-on.

The first leveling technology was developed by Holt Co., a California firm, in 1891. Modern leveling came into being with the invention and patent of a level sensitive mercury switch system invented by Raymond Alvah Hanson in 1946. Raymond's son, Raymond, Jr., produced leveling systems exclusively for John Deere combines until 1995 as R. A. Hanson Company, Inc. In 1995, his son, Richard, purchased the company from his father and renamed it RAHCO International, Inc. In March 2011, the company was renamed Hanson Worldwide, LLC. Production continues to this day.

Hillside leveling has several advantages. Primary among them is an increased threshing efficiency on hillsides. Without leveling, grain and chaff slide to one side of separator and come through the machine in a large ball rather than being separated, dumping large amounts of grain on the ground. By keeping the machinery level, the straw-walker is able to operate more efficiently, making for more efficient threshing. IH produced the 453 combine which leveled both side-to-side and front-to-back, enabling efficient threshing whether on a hillside or climbing a hill head on.

Secondarily, leveling changes a combine`s center of gravity relative to the hill and allows the combine to harvest along the contour of a hill without tipping, a very real danger on the steeper slopes of the region; it is not uncommon for combines to roll on extremely steep hills.

Newer leveling systems do not have as much tilt as the older ones. A John Deere 9600 combine equipped with a Rahco hillside conversion kit will level over to 44%, while the newer STS combines will only go to 35%. These modern combines use the rotary grain separator which makes leveling less critical. Most combines on the Palouse have dual drive wheels on each side to stabilize them.

A leveling system was developed in Europe by the Italian combine manufacturer Laverda which still produces it today.

Sidehill Leveling

Sidehill combines are very similar to hillside combines in that they level the combine to the ground so that the threshing can be efficiently conducted; however, they have some very distinct differences. Modern hillside combines level around 35% on average, older machines were closer to 50%. Sidehill combines only level to 18%. They are sparsely used in the Palouse region. Rather, they are used on the gentle rolling slopes of the mid-west. Sidehill combines are much more mass-produced than their hillside counterparts. The height of a sidehill machine is the same height as a level-land combine. Hillside combines have added steel that sets them up approximately 2–5 feet higher than a level-land combine and provide a smooth ride.

Maintaining Threshing Speed

Another technology that is sometimes used on combines is a continuously variable transmission. This allows the ground speed of the machine to be varied while maintaining a constant engine and threshing speed. It is desirable to keep the threshing speed constant since the machine will typically have been adjusted to operate best at a certain speed.

Allis-Chalmers GLEANER L2.

Self-propelled combines started with standard manual transmissions that provided one speed based on input rpm. Deficiencies were noted and in the early 1950s combines were equipped with what John Deere called the "Variable Speed Drive". This was simply a variable width sheave controlled by spring and hydraulic pressures. This sheave was attached to the input shaft of the transmission. A standard 4 speed manual transmission was still used in this drive system. The operator would select a gear, typically 3rd. An extra control was provided to the operator to allow him to speed up and slow down the machine within the limits provided by the variable speed drive system. By decreasing the width of the sheave on the input shaft of the transmission, the belt would ride higher in the groove. This slowed the rotating speed on the input shaft of the transmission, thus slowing the ground speed for that gear. A clutch was still provided to allow the operator to stop the machine and change transmission gears.

Later, as hydraulic technology improved, hydrostatic transmissions were introduced by Versatile Mfg for use on swathers but later this technology was applied to combines as well. This drive retained the 4 speed manual transmission as before, but this time used a system of hydraulic pumps and motors to drive the input shaft of the transmission. This system is called a Hydrostatic drive system. The engine turns the hydraulic pump capable of pressures up to 4,000 psi (30 MPa). This pressure is then directed to the hydraulic motor that is connected to the input shaft of the transmission. The operator is provided with a lever in the cab that allows for the control of the hydraulic motor's ability to use the energy provided by the pump. By adjusting the swash plate in the motor, the stroke of its pistons are changed. If the swash plate is set to neutral, the pistons do not move in their bores and no rotation is allowed, thus the machine does not move. By moving the lever, the swash plate moves its attached pistons forward, thus allowing them to move within the bore and causing the motor to turn. This provides an infinitely variable speed control from 0 ground speed to what ever the maximum speed is allowed by the gear selection of the transmission. The standard clutch was removed from this drive system as it was no longer needed.

Most if not all modern combines are equipped with hydrostatic drives. These are larger versions of the same system used in consumer and commercial lawn mowers that most are familiar with today. In fact, it was the downsizing of the combine drive system that placed these drive systems into mowers and other machines.

The Threshing Process

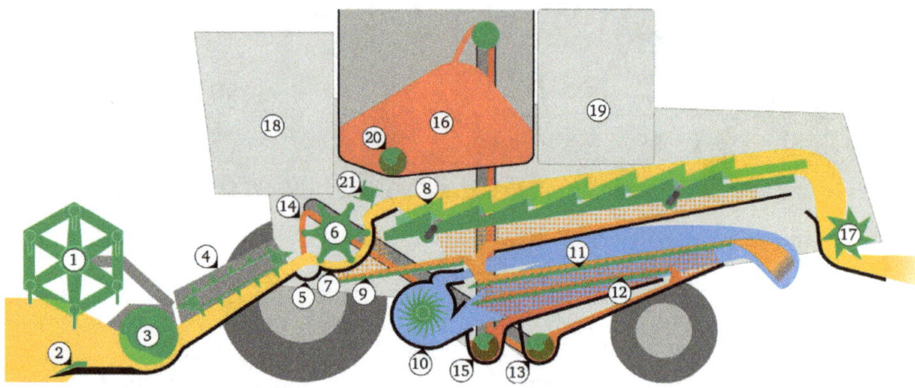

Conventional combine harvester (cut)

1) Reel
2) Cutter bar
3) Header auger
4) Grain conveyor
5) Stone trap
6) Threshing drum
7) Concave
8) Straw walker
9) Grain pan
10) Fan 11) Top Adjustable sieve
12) Bottom sieve
13) Tailings conveyor
14) Rethreshing of tailings
15) Grain auger
16) Grain tank

17) Straw chopper
18) Driver's cab
19) Engine
20) Unloading auger
21) Impeller

First, the header, described above, cuts the crop and feeds it into the threshing cylinder. This consists of a series of horizontal *rasp bars* fixed across the path of the crop and in the shape of a quarter cylinder. Moving rasp bars or rub bars pull the crop through concaved grates that separate the grain and chaff from the straw. The grain heads fall through the fixed concaves. What happens next is dependent on the type of combine in question. In most modern combines, the grain is transported to the shoe by a set of 2, 3, or 4 (possibly more on the largest machines) augers, set parallel or semi-parallel to the rotor on axial mounted rotors and perpendicular Flow" combines.) In older Gleaner machines, these augers were not present. These combines are unique in that the cylinder and concave is set inside feederhouse instead of in the machine directly behind the feederhouse. Consequently, the material was moved by a "raddle chain" from underneath the concave to the walkers. The clean grain fell between the raddle and the walkers onto the shoe, while the straw, being longer and lighter, floated across onto the walkers to be expelled. On most other older machines, the cylinder was placed higher and farther back in the machine, and the grain moved to the shoe by falling down a "clean grain pan", and the straw "floated" across the concaves to the back of the walkers.

Since the Sperry-New Holland TR70 Twin-Rotor Combine came out in 1975, most manufacturers have combines with rotors in place of conventional cylinders. However, makers have now returned to the market with conventional models alongside their rotary line-up. A rotor is a long, longitudinally mounted rotating cylinder with plates similar to rub bars (except for in the above-mentioned Gleaner rotaries).

There are usually two sieves, one above the other. The sieves and basically a metal frame, that has many rows of "fingers" set reasonably close together. The angle of the fingers is adjustable as to change the clearance and control the size of material passing through. The top is set with more clearance than the bottom as to allow a gradual cleaning action. Setting the concave clearance, fan speed, and sieve size is critical to ensure that the crop is threshed properly, the grain is clean of debris, and that all of the grain entering the machine reaches the grain tank or `hopper`. (Observe, for example, that when travelling uphill the fan speed must be reduced to account for the shallower gradient of the sieves.)

Heavy material, e.g., unthreshed heads, fall off the front of the sieves and are returned to the concave for re-threshing.

The straw walkers are located above the sieves, and also have holes in them. Any grain remaining attached to the straw is shaken off and falls onto the top sieve.

When the straw reaches the end of the walkers it falls out the rear of the combine. It can then be baled for cattle bedding or spread by two rotating straw spreaders with rubber arms. Most modern combines are equipped with a straw spreader.

Rotary and Conventional Designs

IH McCormick 141 Combine ca. 1954-57.

Tr85.

For some time, combine harvesters used the conventional design, which used a rotating cylinder at the front-end which knocked the seeds out of the heads, and then used the rest of the machine to separate the straw from the chaff, and the chaff from the grain. The TR70 from Sperry-New Holland was brought out in 1975 as the first rotary combine. Other manufacturers soon followed, International Harvester with their 'Axial Flow' in 1977 and Gleaner with their N6 in 1979.

In the decades before the widespread adoption of the rotary combine in the late seventies, several inventors had pioneered designs which relied more on centrifugal force for grain separation and less on gravity alone. By the early eighties, most major manufacturers had settled on a "walkerless" design with much larger threshing cylinders to do most of the work. Advantages were faster grain harvesting and gentler treatment of fragile seeds, which were often cracked by the faster rotational speeds of conventional combine threshing cylinders.

It was the disadvantages of the rotary combine (increased power requirements and over-pulverization of the straw by-product) which prompted a resurgence of conventional combines in the late nineties. Perhaps overlooked but nonetheless true, when the large engines that powered the rotary machines were employed in conventional machines, the two types of machines delivered similar production capacities. Also, research was beginning to show that incorporating above-ground crop residue (straw) into the soil is less useful for rebuilding soil fertility than previously believed. This meant that working pulverized straw into the soil became more of a hindrance than a benefit. An increase in feedlot beef production also created a higher demand for straw as fodder. Conventional combines, which use straw walkers, preserve the quality of straw and allow it to be baled and removed from the field.

Combine Fires

Grain combine fires are responsible for millions of dollars of loss each year. Fires usually start near the engine where dust and dry crop debris accumulate. Fires can also start when heat is introduced by bearings or gearboxes that have failed. From 1984 to 2000, 695 major grain combine fires were reported to U.S. local fire departments. Dragging chains to reduce static electricity was one method employed for preventing harvester fires, but the role of static electricity linked to causing harvester fires is yet to be established. The application of the appropriate synthetic greases will reduce the friction experienced at crucial points, i.e. chains, sprockets and gear boxes compared to petroleum based lubricants. Engines with synthetic lubricants will also remain significantly cooler during operation.

CULTIVATOR

A **cultivator** is any of several types of farm implement used for secondary tillage. One sense of the name refers to frames with the **teeth** (also called **shanks**) that pierce the soil as they are dragged through it linearly. Another sense refers to machines that use rotary motion of disks or teeth to accomplish a similar result. The **rotary tiller** is a principle example.

F210 Honda tiller

1949 Farmall C with C-254-A two-row cultivator

A tractor-mounted tiller.

Tines close-up

A cultivator pulled by a tractor in Canada in 1943.

Cultivators stir and pulverize the soil, either before planting (to aerate the soil and prepare a smooth, loose seedbed) or after the crop has begun growing (to kill weeds—controlled disturbance of the topsoil close to the crop plants kills the surrounding weeds by uprooting them, burying their leaves to disrupt their photosynthesis, or a combination of both). Unlike a harrow, which disturbs the entire surface of the soil, cultivators are designed to disturb the soil in careful patterns, sparing the crop plants but disrupting the weeds.

Cultivators of the toothed type are often similar in form to chisel plows, but their goals are different. Cultivator teeth work near the surface, usually for weed control, whereas chisel plow shanks work deep beneath the surface, breaking up hardpan. Consequently, cultivating also takes much less power per shank than does chisel plowing.

Small toothed cultivators pushed or pulled by a single person are used as garden tools for small-scale gardening, such as for the household's own use or for small market gardens. Similarly sized rotary tillers combine the functions of harrow and cultivator into one multipurpose machine.

Cultivators are usually either self-propelled or drawn as an attachment behind either a two-wheel tractor or four-wheel tractor. For two-wheel tractors they are usually rigidly fixed and powered via couplings to the tractors` transmission. For four-wheel tractors they are usually attached by means of a three-point hitch and driven by a power take-off (PTO). Drawbar hookup is also still commonly used worldwide. Draft-animal power is sometimes still used today, being somewhat common in developing nations although rare in more industrialized economies.

The basic idea of soil scratching for weed control is ancient and was done with hoes or mattocks for millennia before cultivators were developed. Cultivators were originally drawn by draft animals (such as horses, mules, or oxen) or were pushed or drawn by people. In modern commercial agriculture, the amount of cultivating done for weed control has been greatly reduced via use of herbicides instead. However, herbicides are not always desirable—for example, in organic farming.

The powered rotary hoe was invented by Arthur Clifford Howard who, in 1912, began experimenting with rotary tillage on his father's farm at Gilgandra, New South Wales, Australia. Initially using his father`s steam tractor engine as a power source, he found that ground could be mechanically tilled without soil-packing occurring, as was the case with normal ploughing. His earliest designs threw the tilled soil sideways, until he improved his invention by designing an L-shaped blade mounted on widely spaced flanges fixed to a small-diameter rotor. With fellow apprentice Everard McCleary, he established a company to make his machine, but plans were interrupted by World War I. In 1919 Howard returned to Australia and resumed his design work, patenting a design with 5 rotary hoe cultivator blades and an internal combustion engine in 1920.

In March 1922, Howard formed the company Austral Auto Cultivators Pty Ltd, which later became known as Howard Auto Cultivators. It was based in Northmead, a suburb of Sydney, from 1927.

Meanwhile, in North America during the 1910s, tractors were evolving away from traction engine-sized monsters toward smaller, lighter, more affordable machines. The Fordson tractor especially had made tractors affordable and practical for small and medium family farms for the first time in history. Cultivating was somewhat of an afterthought in the Fordson`s design, which reflected the fact that even just bringing practical motorized tractive power alone to this market segment was in itself a milestone. This left an opportunity for others to pursue better motorized cultivating. Between 1915 and 1920, various inventors and farm implement companies experimented with a class of machines referred to as *motor cultivators*, which were simply modified horse-drawn shank-type cultivators with motors added for self-propulsion. This class of machines found limited market success. But by 1921 International Harvester had combined motorized cultivating with the other tasks of tractors (tractive power and belt work) to create the Farmall, the general-purpose tractor tailored to cultivating that basically invented the category of row-crop tractors.

In Australia, by the 1930s, Howard was finding it increasingly difficult to meet a growing worldwide demand for exports of his machines. He travelled to the United Kingdom, founding the company Rotary Hoes Ltd in East Horndon, Essex, in July 1938. Branches of this new company subsequently opened in the United States of America, South Africa, Germany, France, Italy, Spain, Brazil, Malaysia, Australia and New Zealand. It later became the holding company for Howard Rotavator Co. Ltd. The Howard Group of companies was acquired by the Danish Thrige Agro Group in 1985, and in December 2000 the Howard Group became a member of Kongskilde Industries of Soroe, Denmark.

When herbicidal weed control was first widely commercialized in the 1950s and 1960s, it played into that era's optimistic worldview in which sciences such as chemistry would usher in a new age of modernity that would leave old-fashioned practices (such as weed control via cultivators) in the dustbin of history. Thus herbicidal weed control was adopted very widely, and in some cases too heavily and hastily. In subsequent decades, people overcame this initial imbalance and came to realize that herbicidal weed control has limitations and externalities, and it must be managed intelligently. It is still widely used, and probably will continue to be indispensable to affordable food production worldwide for the foreseeable future; but its wise management includes seeking alternate methods, such as the traditional standby of mechanical cultivation, where practical.

Industrial Use

To the extent that cultivating is done commercially today (such as in truck farming), it is usually powered by tractors, especially row-crop tractors. Industrial cultivators can vary greatly in size and shape, from 10 feet (3 m) to 80 feet (24 m) wide. Many are equipped with hydraulic wings that fold up to make road travel easier and safer. Different types are used for preparation of fields before planting, and for the control of weeds between row crops. The cultivator may be an implement trailed after the tractor via a drawbar; mounted on the three-point hitch; or mounted on a frame beneath the tractor. Active cultivator implements are driven by a power take-off shaft. While most cultivator are considered a secondary tillage implement, active cultivators are commonly used for primary tillage in lighter soils instead of plowing. The largest versions available are about 6 m (20 ft) wide, and require a tractor with an excess of 150 horsepower (110 kW) (PTO) to drive them.

Field cultivators are used to complete tillage operations in many types of arable crop fields. The main function of the field cultivator is to prepare a proper seedbed for the crop to be planted into, to bury crop residue in the soil (helping to warm the soil before planting), to control weeds, and to mix and incorporate the soil to ensure the growing crop has enough water and nutrients to grow well during the growing season. The implement has many shanks mounted on the underside of a metal frame, and small narrow rods at the rear of the machine that smooth out the soil surface for easier travel later when planting. In most field cultivators, one-to-many hydraulic cylinders raise and lower the implement and control its depth.

Row Crop Cultivators

The main function of the row crop cultivator is weed control between the rows of an established crop. Row crop cultivators are usually raised and lowered by a three-point hitch and the depth is controlled by gauge wheels.

Sometimes referred to as *sweep cultivators*, these commonly have two center blades that cut weeds from the roots near the base of the crop and turn over soil, while two rear sweeps further outward than the center blades deal with the center of the row, and can be anywhere from 1 to 36 rows wide.

Home made sweep. Notice the inner and outer "sweep" blades.

Garden Cultivators

Small tilling equipment, used in small gardens such as household gardens and small commercial gardens, can provide both primary and secondary tillage. For example, a rotary tiller does both the "plowing" and the "harrowing", preparing a smooth, loose seedbed. It does not provide the row-wise weed control that cultivator teeth would. For that task, there are single-person-pushable toothed cultivators.

Variants and Trademarks

A Japanese two-wheel tractor.

Rotary tillers are a type of cultivators. Rotary tillers are popular with home gardeners who want large vegetable gardens. The garden may be tilled a few times before planting each crop. Rotary tillers may be rented from tool rental centers for single-use applications, such as when planting grass.

A small rotary hoe for domestic gardens was known by the trademark Rototiller and another, made by the Howard Group, who produced a range of rotary tillers, was known as the Rotavator.

Rototiller

The small rototiller is typically propelled forward via a (1–5 horsepower or 0.8–3.5 kilowatts) petrol engine rotating the tines, and do not have powered wheels, though they may have small transport/level control wheel(s). To keep the machine from moving forward too fast, an adjustable tine is usually fixed just behind the blades so that through friction with deeper un-tilled soil, it acts as a brake, slowing the machine and allowing it to pulverize the soils. The slower a rototiller moves forward, the more soil tilth can be obtained. The operator can control the amount of friction/braking action by raising and lowering the handlebars of the tiller. Rototillers do not have a reverse as such backwards movement towards the operator could cause serious injury. While operating, the rototiller can be pulled backwards to go over areas that were not pulverized enough, but care must be taken to ensure that the operator does not stumble and pull the rototiller on top of himself. Rototilling is much faster than manual tilling, but notoriously difficult to handle and exhausting work, especially in the heavier and higher horsepower models. If the rototiller's blades catch on unseen subsurface objects, such as tree roots and buried garbage, it can cause the rototiller to abruptly and violently move in any direction.

Rotavator

Unlike the Rototiller, the self-propelled Howard Rotavator is equipped with a gearbox and driven forward, or held back, by its wheels. The gearbox enables the forward speed to be adjusted while the rotational speed of the tines remains constant which enables the operator to easily regulate the extent to which soil is engaged. For a two-wheel tractor rotavator this greatly reduces the workload of the operator as compared to a rototiller. These rotavators are generally more heavy duty, come in higher power (4–18 horsepower or 3–13 kilowatts) with either petrol or diesel engines and can cover much more area per hour. The trademarked word "Rotavator" is one of the longest single-word palindromes in the English language.

Mini tiller

Mini tillers are a new type of small agricultural tillers or cultivators used by farmers or homeowners. These are also known as power tillers or garden tillers. Compact, powerful and, most importantly, inexpensive, these agricultural rotary tillers are providing alternatives to four-wheel tractors and in the small farmers' fields in developing countries are more economical than four-wheel tractors.

Two-wheel tractor

The higher power "riding" rotavators cross out of the home garden category into farming category, especially in Asia, Africa and South America, capable of preparing 1 hectare of land in 8–10 hours. These are also known as *power tillers* or *walking tractors*. Years ago they were considered only useful for rice growing areas, where they were fitted with steel cage-wheels for traction, but now the same are being used in both wetland and dryland farming all over the world. They have multiple functions with related tools for dryland or paddys, pumping, transportation, threshing, ditching, spraying pesticide. They can be used on hills, mountains, in greenhouses and orchards. Diesel designs are more popular in developing countries than gasoline.

BALER

A Claas large round baler

An Abbriata Roto-Baler ejecting a net-wrapped small round bale.

A **baler**, most often called a **hay baler** is a piece of farm machinery used to compress a cut and raked crop (such as hay, cotton, flax straw, salt marsh hay, or silage) into compact bales that are easy to handle, transport, and store. Often bales are configured to dry and preserve some intrinsic (e.g. the nutritional) value of the plants bundled. Several different types of balers are commonly used, each producing a different type of bale – rectangular or cylindrical, of various sizes, bound with twine, strapping, netting, or wire.

Industrial balers are also used in material recycling facilities, primarily for baling metal, plastic, or paper for transport.

Before the 19th century, hay was cut by hand and most typically stored in haystacks using hay forks to rake and gather the scythed grasses into optimal sized heaps — neither too large (promoting conditions that might create spontaneous combustion), nor too small, so much of the pile is susceptible to rotting. These haystacks lifted most of the plant fibers up off the ground, letting air in and water drain out, so the grasses could dry and cure, to retain nutrition for livestock feed at a later time. In the 1860s mechanical cutting devices were developed; from these came the modern devices including mechanical mowers and balers. In 1872 a reaper that used a knotter device to bundle and bind hay was invented by Charles Withington; this was commercialized in 1874 by Cyrus McCormick. In 1936, Innes invented an automatic baler that tied bales with twine using Appleby-type knotters from a John Deere grain binder; an improved version patented by Ed Nolt in 1939 was more reliable and became commonly used.

Round baler

Allis Chalmers Rotobaler.

Round baler dumping a fresh bale.

The most common type of baler in industrialized countries today is the round baler. It produces cylinder-shaped "round" or "rolled" bales. The design has a "thatched roof" effect that withstands weather well.Grass is rolled up inside the baler using rubberized belts, fixed rollers, or a combination of the two. When the bale reaches a predetermined size, either netting or twine is wrapped around it to hold its shape. The back of the baler swings open, and the bale is discharged. The bales are complete at this stage, but they may also be wrapped in plastic sheeting by a bale wrapper, either to keep hay dry when stored outside or convert damp grass into silage. Variable-chamber large round balers typically produce bales from 48 to 72 inches (120 to 180 cm) in diameter and up to 60 inches (150 cm) in width. The bales can weigh anywhere from 1,100 to 2,200 pounds (500 to 1,000 kg), depending upon size, material, and moisture content. Common modern small round balers (also called "mini round balers" or "roto-balers") produce bales 20 to 22 inches (51 to 56 cm) in diameter and 20.5 to 28 inches (52 to 71 cm) in width, generally weighing from 40 to 55 pounds (18 to 25 kg).

Originally conceived by Ummo Luebben circa 1910, the first round baler did not see production until 1947 when Allis-Chalmers introduced the Roto-Baler. Marketed for the water-shedding and light weight properties of its hay bales, AC had sold nearly 70,000 units by the end of production in 1960. The next major innovation began in 1965 when a graduate student at Iowa State University, Virgil Haverdink, sought out Wesley F. Buchele, a professor of Agricultural Engineering, seeking a research topic for a master thesis. Over the next year Buchele and Haverdink developed a new design for a large round baler, completed and tested in 1966, and thereafter dubbed the Buchele-Haverdink large round baler.The large round bales were about 5 ft in diameter, 7 ft long, and they weighed about 600 lb after they dried—about 5 lb/ft^3. The design was promoted as a "Whale of a Bale" and Iowa State University now explains the innovative design as follows:

"Farmers were saved from the backbreaking chore of slinging hay bales in the 1960s when Iowa State agricultural engineering professor Wesley Buchele and a group of student researchers invented a baler that produced large, round bales that could be moved by tractor. The baler has become the predominant forage-handling machine in the United States."

In the summer of 1969, the Australian Econ Fodder Roller baler came out, a design that made a 300-lb ground-rolled bale. In September of that same year The Hawkbilt Company of Vinton, Iowa, contacted Dr. Buchele about his design, then fabricated a large ground-rolling round baler which baled hay that had been laid out in a windrow, and began manufacturing large round balers in 1970. In 1972, Gary Vermeer of Pella, Iowa, designed and fabricated a round baler after the design of the A-C Roto-Baler, and the Vermeer Company began selling its model 605 - the first modern round baler. The Vermeer design used belts to compact hay into a cylindrical shape as is seen today. In the early 1980s, collaboration between Walterscheid and Vermeer produced the first effective uses of CV joints in balers, and later in other farm machinery. Due to the heavy torque required for such equipment, double Cardan joints are primarily used. Former Walterscheid engineer Martin Brown is credited with "inventing" this use for universal joints.

By 1975, fifteen American and Canadian companies were manufacturing large round balers.

Transport, Handling, and Feeding

Short-haul Transport and on-field Handling

A large round bale.

Due to the ability for round bales to roll away on a slope, they require specific treatment for safe transport and handling. Small round bales can typically be moved by hand or with lower-powered equipment. Large round bales, due to their size and weight (they can weigh a ton or more) require special transport and moving equipment.

The most important tool for large round bale handling is the bale spear or spike, which is usually mounted on the back of a tractor or the front of a skid-steer. It is inserted into the approximate center of the round bale, then lifted and the bale is hauled away. Once at the destination, the bale is set down, and the spear pulled out. Careful placement of the spear in the center is needed or the bale can spin around and touch the ground while in transport, causing a loss of control. When used for wrapped bales that are to be stored further, the spear makes a hole in the wrapping that must be sealed with plastic tape to maintain a hermetic seal.

Alternatively, a grapple fork may be used to lift and transport large round bales. The grapple fork is a hydraulically driven implement attached to the end of a tractor's bucket loader. When the hydraulic cylinder is extended, the fork clamps downward toward the bucket, much like a closing hand. To move a large round bale, the tractor approaches the bale from the side and places the bucket underneath the bale. The fork is then clamped down across the top of the bale, and the bucket is lifted with the bale in tow. Grab hooks installed on the bucket of a tractor are another tool used to handle round bales, and be done by a farmer with welding skills by welding two hooks and a heavy chain to the outside top of a tractor front loader bucket.

Long-haul Transport

The rounded surface of round bales poses a challenge for long-haul, flat-bed transport, as they could roll off of the flat surface if not properly supported. This is particularly the case with large round bales; their size makes them difficult to flip, so it may not be feasible to flip many of them onto the flat surface for transport and then re-position them on the round surface at the destination. One option that works with both large and small round bales is to equip the flat-bed trailer with guard-rails at either end, which prevent bales from rolling either forward or backward. Another solution is the saddle wagon, which has closely spaced rounded saddles or support posts in which round bales sit. The tall sides of each saddle prevent the bales from rolling around while on the wagon, as the bale settles down in between posts. On 3 September 2010, on the A381 in Halwell near Totnes, Devon, UK an early member of British rock group ELO Mike Edwards was killed when his van was crushed by a large round bale. The cellist, 62, died instantly when the 600 kilograms (1,300 lb) bale fell from a tractor on nearby farmland before rolling onto the road and crushing his van.

Feeding

A large round bale can be directly used for feeding animals by placing it in a feeding area, tipping it over, removing the bale wrap, and placing a protective ring (a *ring feeder*) around the outside so that animals don't walk on hay that has been peeled off the outer perimeter of the bale. The round baler's rotational forming and compaction process also enables both large and small round bales to be fed out by unrolling the bale, leaving a continuous flat strip in the field or behind a feeding barrier.

Silage or Haylage

A recent innovation in hay storage has been the development of the silage or haylage bale, which is a high-moisture bale wrapped in plastic film. These are baled much wetter than hay bales, and are usually smaller than hay bales because the greater moisture content makes them heavier and harder to handle. These bales begin to ferment almost immediately, and the metal bale spear stabbed into the core becomes very warm to the touch from the fermentation process.

Silage or haylage bales may be wrapped by placing them on a rotating bale spear mounted on the rear of a tractor. As the bale spins, a layer of plastic cling film is applied to the exterior of the bale. This roll of plastic is mounted in a sliding shuttle on a steel arm and can move parallel to the bale axis, so the operator does not need to hold up the heavy roll of plastic. The plastic layer extends over the ends of the bale to form a ring of plastic approximately 12 inches (30 cm) wide on the ends, with hay exposed in the center.

To stretch the cling-wrap plastic tightly over the bale, the tension is actively adjusted with a knob on the end of the roll, which squeezes the ends of the roll in the shuttle. In this example wrapping video, the operator is attempting to use high tension to get a flat, smooth seal on the right end. However, the tension increases too much and the plastic tears off. The operator recovers by quickly loosening the tension and allows the plastic to feed out halfway around the bale before reapplying the tension to the sheeting.

These bales are placed in a long continuous row, with each wrapped bale pressed firmly against all the other bales in the row before being set down onto the ground. The plastic wrap on the ends of each bale sticks together to seal out air and moisture, protecting the silage from the elements. The end-bales are hand-sealed with strips of cling plastic across the opening.

The airtight seal between each bale permits the row of round bales to ferment as if they were in a silo bag, but they are easier to handle than a silo bag, as they are more robust and compact. The plastic usage is relatively high, and there is no way to reuse the silage-contaminated plastic sheeting, although it can be recycled or used as a fuel source via incineration. The wrapping cost is approximately US$5 per bale.

An alternative form of wrapping uses the same type of bale placed on a bale wrapper, consisting of pair of rollers on a turntable mounted on the three-point linkage of a tractor. It is then spun about two axes while being wrapped in several layers of cling-wrap plastic film. This covers the ends and sides of the bale in one operation, thus sealing it separately from other bales. The bales are then moved or stacked using a special pincer attachment on the front loader of a tractor, which does not damage the film seal. They can also be moved using a standard bale spike, but this punctures the airtight seal, and the hole in the film must be repaired after each move.

Plastic-wrapped bales must be unwrapped before being fed to livestock to prevent accidental ingestion of the plastic. Like round hay bales, silage bales are usually fed using a *ring feeder*.

Large rectangular baler.

Large Rectangular Baler

Another type of baler in common use, in some areas, will produce large rectangular bales, each bound with a half dozen or so strings of twine, which are then knotted. Such bales are highly compacted and generally weigh somewhat more than round bales. The large rectangular bales are several times larger than the similar small bales. In the prairies of Canada, the large rectangular balers are also called "prairie raptors".

Rectangular Bale Handling and Transport

Large rectangular bales in a field, Charente, France. Sizes of stacks of baled hay need carefully managed, as the curing process is exothermic and the built up heat around internal bales can reach ignition temperatures in the right weather history and atmospheric conditions. Building a deep stack either too wide, or too high increases the risk of spontaneous ignition.

Rectangular bales are easier to transport than round bales, since there is little risk of the bale rolling off the back of a flatbed trailer. The rectangular shape also saves space and allows a complete solid slab of hay to be stacked for transport and storage. Most balers allow adjustment of length and it is common to produce bales of twice the width, allowing stacks with brick-like alternating groups overlapping the row below at right angles, creating a strong structure.

They are well-suited for large scale livestock feedlot operations, where many tons of feed are rationed every hour. Most often, they are baled small enough that one person can carry or toss them where needed.

Due to the huge rectangular shape, large spear forks, or squeeze grips are mounted to heavy lifting machinery, such as large fork lifts, tractors equipped with front end loaders, telehandlers, hay squeezes or wheel loaders, to lift these bales.

Small Rectangular Baler

A small square baler.

A type of baler that produces small rectangular (often called "square") bales was once the most prevalent form of baler, but is less common today. It is primarily used on small acreages where large equipment is impractical, and also for the production of hay for small operations, particularly horse owners who may not have access to the specialized feeding machinery used for larger baled. Each bale is about 15 in x 18 in x 40 in (40 x 45 x 100 cm). The bales are usually wrapped with two, but sometimes three, or more strands of knotted twine. The bales are light enough for one person to handle, about 45 to 60 pounds (20 to 27 kg).

To form the bale, the material to be baled, (which is often hay or straw) in the windrow is lifted by tines in the baler's *reel*. This material is then *packed* into the bale chamber, which runs the length of one side of the baler (normally the right hand side when viewed from the front). A combination plunger and knife move back and forth in the front of this chamber, with the knife closing the door into the bale chamber as it moves backwards. The plunger and knife are attached to a heavy asymmetrical flywheel to provide extra force as they pack the bales. A measuring device—normally a spiked wheel that is turned by the emerging bales—measures the amount of material that is being compressed and, at the appropriate length it triggers the *knotters* that wrap the twine around the bale and tie it off. As the next bale is formed the tied one is driven out of the rear of the baling chamber onto the ground or onto a special wagon or collecting sled hooked to the rear of the baler. This process continues as long as there is material to be baled, and twine to tie it with. The bales emerge with four sides. The twine runs, in two parallel loops, around the wider sides. Of the two narrower sides, there is a cut side and a dull side, and when stacked for storage or transport the bales are normally positioned with the cut side facing outwards.

This form of bale is not much used in large-scale commercial agriculture, because of the costs involved in handling many small bales. However, it enjoys some popularity in small-scale, low-mechanization agriculture and horse-keeping. Besides using simpler machinery and being easy to handle, these small bales can also be used for insulation and building materials in straw-bale construction. Square bales may generally weather better than round bales because a more much dense stack can be put up. However, they don`t shed water as round bales do. Convenience is also a major factor in farmers deciding to continue putting up square bales, as they make feeding and bedding in confined areas (stables, barns, etc.) much easier.

Many of these older balers are still to be found on farms today, particularly in dry areas, where bales can be left outside for long periods.

The automatic-baler for small square bales took on most of its present form in 1940. It was first manufactured by the *New Holland Ag* and it used a small petrol engine to provide operating power. It is based on a 1937 invention for a twine-tie baler with automatic pickup.

Wire Balers

Stationary baler.

Bales prior to 1937 were manually wire-tied with two baling wires. Even earlier, the baler was a stationary implement, driven with a tractor or stationary engine using a belt on a belt pulley, with the hay being brought to the baler and fed in by hand. Later, balers were made mobile, with a 'pickup' to gather up the hay and feed it into the chamber. These often used air cooled gasoline engines mounted on the baler for power. The biggest change to this type of baler since 1940 is being powered by the tractor through its power take-off (PTO), instead of by a built-in internal combustion engine.

In present-day production, small square balers can be ordered with twine knotters or wire tie knotters.

Not all stationary wire tying balers used 2 wires. It was not uncommon for the larger bale size (usually 17" x 22") machines to use `boards` that had three slots for wires and hence tied three wires per bale. Most North American manufacturers produced these machines as either regular models or as size options. `Small square` three wire tying pick-up balers were available from the early 1930s, principally from J. I. Case & Co. and Ann Arbor. These machines were hand tying and hand threading machines. Although New Holland credits itself with inventing the `successful small square twine tying baler`, it produced such machines for the first time in 1940 after acquiring Ed Nolte and his baler. This baler baled successfully from 1937 onwards. Certainly the quality of the New Holland machines, popularised twine tying hay balers. In Europe, in as early as 1939, both Claas of Germany and Rousseau SA of France had automatic twine tying pick-up balers. Most of these produced low density bales though. The first successful pick-up balers were made by the Ann Arbor Company in 1929. Ann Arbor were acquired by the Oliver Farm Equipment Company in 1943. Despite their head start on the rest of the field, no Ann Arbor balers carried automatic knotters or twisters. Oliver introduced these in 1949.

Square/wire Bale History

Pickup and Handling Methods

In the 1940s most farmers would bale hay in the field with a small tractor with 20 or less horsepower, and the tied bales would be dropped onto the ground as the baler moved through the field. Another team of workers with horses and a flatbed wagon would come by and use a sharp metal hook to grab the bale and throw it up onto the wagon while an assistant stacks the bale, for transport to the barn.

A later time-saving innovation was to tow the flatbed wagon directly behind the baler, and the bale would be pushed up a ramp to a waiting attendant on the wagon. The attendant hooks the bale off the ramp and stacks it on the wagon, while waiting for the next bale to be produced.

Eventually, as tractor horsepower increased, the thrower-baler became possible, which eliminated the need for someone to stand on the wagon and pick up the finished bales. The first thrower mechanism used two fast-moving friction belts to grab finished bales and throw them at an angle up in the air onto the bale wagon. The bale wagon was modified from a flatbed into a three-sided skeleton frame open at the front, to act as a catcher's net for the thrown bales.

As tractor horsepower further increased, the next innovation of the thrower-baler was the hydraulic tossing baler. This employs a flat pan behind the bale knotter. As bales advance out the back of the baler, they are pushed onto the pan one at a time. When the bale has moved fully onto the pan, the pan suddenly pops up, pushed by a large hydraulic cylinder, and tosses the bale up into the wagon like a catapult.

The pan-thrower method puts much less stress on the bales compared to the belt-thrower. The friction belts of the belt-thrower stress the twine and knots as they grip the bale, and would occasionally cause bales to break apart in the thrower or when the bales landed in the wagon.

Square bale stacker.

A smaller type of stacker.

Bales may be picked up from the field and stacked using a self-powered machine called a *bale stacker*, *bale wagon* or *harobed*. There are several designs and sizes. One type picks up square bales are dropped by the baler with the strings facing upward. The stacker will drive up to each bale, pick it up and set it on a three-bale-wide table (the strings are now facing upwards). Once three bales are on the table, the table lifts up and back, causing the three bales to face strings to the side again; this happens three more times until there are 16 bales on the main table. This table will lift like the smaller one, and the bales will be up against a vertical table. The machine will hold 160 bales (ten tiers); usually there will be cross-tiers near the center to keep the stack from swaying or collapsing if any weight is applied to the top of the stack. The full load will be transported to a barn; the whole rear of the stacker will tilt upwards until it is vertical. There will be two pushers that will extend through the machine and hold the bottom of the stack from being pulled out from the stacker while it is driven out of the barn.

In Britain (if small square bales are still to be used), they are usually collected as they fall out of the baler in a *bale sledge* dragged behind the baler. This has four channels, controlled by automatic mechanical balances, catches and springs, which sort each bale into its place in a square *eight*. When the sledge is full, a catch is tripped automatically, and a door at the rear opens to leave the eight lying neatly together on the ground. These may be picked up individually and loaded by hand, or they may be picked up all eight together by a *bale grab* on a tractor, a special front loader consisting of many hydraulically powered downward-pointing curved spikes. The square eight will then be stacked, either on a trailer for transport, or in a roughly cubic field stack eight or ten layers high. This cube may then be transported by a large machine attached to the three-point hitch behind a tractor, which clamps the sides of the cube and lifts it bodily.

Storage Methods

Before electrification occurred in rural parts of the United States in the 1940s, some small dairy farms would have tractors but not electric power. Often just one neighbor who could afford a tractor would do all the baling for surrounding farmers still using horses.

To get the bales up into the hayloft, a pulley system ran on a track along the peak of the barn`s hayloft. This track also stuck a few feet out the end of the loft, with a large access door under the track. On the bottom of the pulley system was a bale spear, which is pointed on the end and has retractable retention spikes.

A flatbed wagon would pull up next to the barn underneath the end of the track, the spear lowered down to the wagon, and speared into a single bale. The pulley rope would be used to manually lift the bale up until it could enter the mow through the door, then moved along the track into the barn and finally released for manual stacking in tight rows across the floor of the loft. As the stack filled the loft, the bales would be lifted higher and higher with the pulleys until the hay was stacked all the way up to the peak.

When electricity arrived, the bale spear, pulley and track system were replaced by long motorized bale conveyors known as hay elevators. A typical elevator is an open skeletal frame, with a chain that has dull 3-inch (76 mm) spikes every few feet along the chain to grab bales and drag them along. One elevator replaced the spear track and ran the entire length of the peak of the barn. A second elevator was either installed at a 30-degree slope on the side of the barn to lift bales up to the peak elevator, or used dual front-back chains surrounding the bale to lift bales straight up the side of the barn to the peak elevator.

A bale wagon pulled up next to the lifting elevator, and a farm worker placed bales one at a time onto the angled track. Once bales arrived at the peak elevator, adjustable tipping gates along the length of the peak elevator were opened by pulling a cable from the floor of the hayloft, so that bales tipped off the elevator and dropped down to the floor in different areas of the loft. This permitted a single elevator to transport hay to one part of a loft and straw to another part.

This complete hay elevator lifting, transport, and dropping system reduced bale storage labor to a single person, who simply pulls up with a wagon, turns on the elevators and starts placing bales on it, occasionally checking to make sure that bales are falling in the right locations in the loft.

The neat stacking of bales in the loft is often sacrificed for the speed of just letting them fall and roll down the growing pile in the loft, and changing the elevator gates to fill in open areas around the loose pile. But if desired, the loose bale pile dropped by the elevator could be rearranged into orderly rows between wagon loads.

Usage once in the Barn

The process of retrieving bales from a hayloft has stayed relatively unchanged from the beginning of baling. Typically workers were sent up into the loft, to climb up onto the bale stack, pull bales off the stack, and throw or roll them down the stack to the open floor of the loft. Once the bale is down on the floor, workers climb down the stack, open a cover over a bale chute in the floor of the loft, and push the bales down the chute to the livestock area of the barn.

Most barns were equipped with several chutes along the sides and in the center of the loft floor. This permitted bales to be dropped into the area where they were to be used. Hay bales would be dropped through side chutes, to be broken up and fed to the cattle. Straw bales would be dropped down the center chute, to be distributed as bedding in the livestock standing/resting areas.

Traditionally multiple bales were dropped down to the livestock floor and the twine removed by hand. After drying and being stored under tons of pressure in the haystack, most bales are tightly compacted and need to be torn apart and fluffed up for use.

One recent method of speeding up all this manual bale handling is the bale shredder, which is a large vertical drum with rotary cutting/ripping teeth at the base of the drum. The shredder is placed under the chute and several bales dropped in. A worker then pushes the shredder along the barn aisle as it rips up a bale and spews it out in a continuous fluffy stream of material.

Field of hay bales, lines in field made by baler

Industrial Balers

A specialized baler designed to compact stretch wrap.

Industrial balers are typically used to compact similar types of waste, such as office paper, Corrugated fiberboard, plastic, foil and cans, for sale to recycling companies. These balers are made of steel with a hydraulic ram to compress the material loaded. Some balers are simple and labor-intensive, but are suitable for smaller volumes. Other balers are very complex and automated, and are used where large quantities of waste are handled.

BEAVERSLIDE

A beaverslide with a full stack of hay.

A **beaverslide** is a device for stacking hay made of wooden poles and planks that builds haystacks of loose, unbaled hay stored outdoors to be used as fodder for livestock. The beaverslide consists of a frame supporting an inclined plane up which a load of hay is pushed to a height of about 30 feet (9 m), before dropping through a large gap. The resulting loaf-shaped haystacks can be up to 30 feet high, can weigh up to 20 tons, and can theoretically last up to five or six years. It was invented in the early 1900s and was first called the Beaverhead County Slide Stacker after its place of origin, the Big Hole Valley in Beaverhead County, Montana. The name was quickly shortened to "beaverslide."

Early settlers in the American west initially stored hay for their livestock under shelter in barns and haylofts. However, unlike the east, where hay is fed as a supplemental form of forage, the northern plains had lengthy and severe winter weather and therefore large quantities of hay were needed to provide adequate forage for animals. Most haylofts were insufficient to store the quantities needed, but in the arid western United States, unlike the more humid east, hay could be stored without the protection of a barn. As a result, settlers used a variety of methods to stack and store large amounts of hay, inventing a number of agricultural machines to lift hay including hay derricks and various slides, including a predecessor to the beaverslide called a ram stacker.

About 1908 the beaverslide was invented in Montana by two ranchers, Herbert S. Armitage and David J. Stephens, who ranched near Briston, in the Big Hole Valley, of Beaverhead County in southwestern Montana. Armitage and Stephens filed for a patent on September 7, 1909 and it was awarded on May 31, 1910. The beaverslide may have been called the "Sunny Slope Slide Stacker" at one time, but that name does not appear in the patent. Armitage and Stephens themselves referred to it as the "Beaverhead County Slide Stacker", which quickly became just "beaverslide".

The beaverslide was somewhat mobile, inexpensive, handled large amounts of hay, and was easily built. It was faster to use than early balers and made windproof haystacks. It rapidly gained popularity in southwestern Montana and adjacent parts of eastern Idaho, with its use spreading to other western states and Canada in places where light meadow grass was put up as hay. In regions where it had been adopted it remained in common use into the 1990s. While use of a beaverslide is labor-intensive, and it has not been commonly used in the 21st century, some ranchers are returning to it to save fuel costs. Still others never abandoned it because of the large cash outlay required to purchase modern mechanized balers.

Construction

View of beaverslide without haystack.

The beaverslide is constructed of a rigid pole frame in the form of a right-angle triangle that supports a steeply inclined, slatted, plank ramp, with or without sides, approximately 50–60 feet (15–18 m) long. The size and angle of individual beaverslides varies greatly and reflects local needs. Beaverslides were originally of all-wooden construction, usually lodgepole pine, and could last 10 to 15 years. In the 1970s, some components began to be made of metal, which are longer-lasting. The inclined ramp, about 15–20 feet (5–6 m) wide, is made of smooth wooden or metal slats and is about two thirds the length of the poles. In the 1920s it became possible to extend the height of the slide so that the hay could be thrown further. In the 1950s movable wings were placed on either side of the ramp so that the hay could be stacked more neatly. A flat, toothed wooden platform called a "basket" or rack is suspended by a system of cables and pulleys from the poles. It is raised to bring hay to the top of the slide, and then lowered back down along the length of the ramp. A backstop, usually an open wooden grid held up by poles, is set at the far end of the stack helps hold the completed stack in place. A fence of wood panels or other materials is often placed around the stack to keep out livestock.

Use

Haystack created by a beaver slide.

A beaverslide will raise hay to a height that allows a haystack to be built as much as 30 feet high. A large hay crew is required, with a minimum of six people to operate all components. A load of hay is delivered to the base of the beaverslide, often pushed by a buckrake drawn by a team of horses or a tractor. The hay is loaded onto the rack, which when full is drawn up the inclined ramp by cables powered either by a second team of horses or a motorized vehicle such as a pickup or a tractor. At the top of the incline, the hay falls onto the stack and the rack is lowered for another load. The term "butt" describes the hay stacked by the beaverslide and has two meanings. A "butt" can be the amount of hay on a fully loaded rack, but the term also refers to the amount of hay that can be stacked by the beaverslide without moving it, roughly 24 tons of hay. The hay at the top of each haystack is stomped and piled higher towards the middle to allow rain to run off. Depending on the size of the field and the amount of hay produced per acre, once a beaverslide has created a stack, it can be moved a few feet to make a long, continuous haystack, or moved a longer distance to create multiple stacks within a field. Many are built on skids to facilitate being moved from field to field. If the hay is stacked properly, and remains uneaten, the hay in a beaverslide-constructed stack remains good at least two to three years, with some ranchers claiming it could last up to five or six years. In contrast, baled hay stored outdoors can begin to go bad after only one year.

REFERENCES

1. Agricultural engineering Britannica Online. Retrieved 2012-12-25.
2. Tractor (vehicle) Britannica Online. Retrieved 2012-12-25.

3. http://www.wired.com/2015/02/new-high-tech-farm-equipment-nightmare-farmers/
4. http://www.wired.com/2015/04/dmca-ownership-john-deere/
5. A worldwide community of farmers that build and modify our own tools. http://farmhack.org/app/
6. Open source CNC farming http://go.farmbot.it/
7. Pearce, J.M.(2015). Applications of Open Source 3-D Printing on Small Farms. Organic Farming 1(1), 19-35. DOI: 10.12924/of2014.01010019
8. Exemption to Prohibition on Circumvention of Copyright Protection Systems for Access Control Technologies http://copyright.gov/1201/2015/fedreg-publicinspectionFR.pdf
9. Ernst, Lisa; Swaney, Alexandra. "The Beaverslide: Homegrown Haying Technology". Folklife. Montana Arts Council. Retrieved September 28, 2012.
10. "Family Stacks Hay The Old Way". High Plains Journal. September 10, 2007. Retrieved September 28, 2012.
11. Kerson, Mike (December 5, 2010). Horse-Drawn Loose Haying at Grant-Kohrs Ranch — Deer Lodge, Montana. Video Mike Productions. Event occurs at 0:00–0:17, 0:22–0:42. Retrieved September 28, 2012.
12. Klinkenborg, Verlyn (1997). Making Hay. Guilford, CT: The Lyons Press. ISBN 978-1-58574-726-9.
13. Merrill-Maker, Andrea (2006). Montana Almanac (2 ed.). Kearney, NE: Morris Book Publishing. ISBN 978-0-7627-3655-3.
14. Munday, Pat (2001). "6". Montana`s Last Best River: The Big Hole and Its People. Guilford, CT: The Lyons Press. p. 75. ISBN 978-1-58574-331-5.
15. "Grant-Kohrs Ranch". National Park Service. Retrieved September 28, 2012.
16. Oberling, Nicholas (Winter 2007). "The Big Hole River In Paint: a Benefit for the Big Hole River Foundation" (PDF). Big Hole River Foundation. Archived from the original (PDF) on May 15, 2008. Retrieved September 28, 2012.
17. Sando, Linnea Christiana (2010). Landscape Imprints of Haying Technology in Eastern Idaho and Western Montana (PDF) (M.A.). Kansas State University. Retrieved October 1, 2012.
18. US Patent 959906, Herbert S. Armitage & David J. Stephens, "Hay-Stacker", issued May 31, 1910. U.S. Patent 959,906.
19. Welsch, Jeff; Moore, Sherry (2011). Backroads & Byways of Montana: Drives, Day Trips & Weekend Excursions. Woodstock, VT: The Countryman Press. ISBN 978-0-88150-899-4.

CHAPTER 2

Water Use and Land Use

Chapter Contents

- **Introduction**
- **Water Footprint**
- **Calculation for Different Actors**
- **Life Cycle Analysis of Water Use**
- **Water Footprint of Products**
- **Water Footprint of Individual Consumers**
- **Water Footprint of Companies**
- **Water Footprint of Nations**
- **Criticism of Water Footprint and Virtual Water**
- **Environmental Water Use**
- **Sectoral Distributions of Withdrawn Water Use**
- **Land Use**
- **Cultural Landscape**

Water consumption is simply the use of water in any form for any reason. While most people tend to think of water consumption as nothing more than drinking, water is consumed in a variety of ways, with human consumption only making up a small fraction of the use. It is used in irrigation, cleaning, cooking, and even as a coolant, in some cases.

INTRODUCTION

Water use can mean the amount of water used for the production of a given quantity of some product or crop, or the amount allocated for a particular purpose.

Globally, of precipitation falling on land each year (about 117,000 km^3 (28,000 cu mi)), about 4 percent is used by rainfed agriculture and about half is subject to evaporation and transpiration in forests and other natural or quasi-natural landscapes. The remainder, which goes to groundwater replenishment and surface runoff, is sometimes called "total actual renewable freshwater resources". Its magnitude was recently estimated at 52,579 km^3 (12,614 cu mi)/year. It represents water that can be used either instream or after withdrawal from surface and groundwater sources. Of this remainder, about 3,918 km^3

were withdrawn in 2007, of which 2,722 km^3 (69 percent) were used by agriculture, and 734 km^3 (19 percent) by other industry. Most agricultural use of withdrawn water is for irrigation, which uses about 5.1 percent of total actual renewable freshwater resources. World water use has been growing rapidly in the last hundred years.

There are numerous measures of water use, including total water use, drinking water consumption, non-consumptive use, withdrawn water use (from surface and groundwater sources), instream use, water footprint, etc. Each of these (and other) measures of water use is appropriate for some purposes and inappropriate for others. Water "footprints" have become popular measures of use, e.g. in relation to personal consumption. The term "water footprint" is often used to refer to the amount of water used by an individual, community, business, or nation, or the amount of water use associated with (although not necessarily assignable to) a product.

The total water footprint of a typical 3-person household in the U.S. is 23,360 l (5,140 imp gal; 6,170 US gal). By comparison, a typical single family home in the U.S. only uses about 262 l (58 imp gal; 69 US gal) of water per day (2008 estimate). This includes (in decreasing order) toilet use, washing machine use, showers, baths, faucet use, and leaks.

WATER FOOTPRINT

The **water footprint** of an individual, community or business is defined as the total volume of freshwater used to produce the goods and services consumed by the individual or community or produced by the business. Water use is measured in water volume consumed (evaporated) and/or polluted per unit of time. A water footprint can be calculated for any well-defined group of consumers (e.g., an individual, family, village, city, province, state or nation) or producers (e.g., a public organization, private enterprise or economic sector). The water footprint is a geographically explicit indicator, not only showing volumes of water use and pollution, but also the locations.

The water footprint of a country is related to what its people eat. For example, it is a common thought that the water involved in a cup of coffee is just the water in the cup. There is actually 140 litres of water involved. The 140 litres of water is the amount of water that was used to grow, produce, package, and ship the coffee beans. A hamburger needs an estimated 2,400 litres of water. This hidden water is technically called virtual water. Therefore, eating a lot of meat means a large water footprint. However, care is needed to avoid misunderstanding the significance of water footprints of food. (See Water footprint of products, below.)

History

The water footprint concept was introduced in 2002 by Arjen Y. Hoekstra from UNESCO-IHE as an alternative indicator of water use. Water footprint is one of a family of footprint indicators, which also includes carbon footprint and land footprint. The water footprint concept is further related to the idea of virtual water trade introduced by Professor John Allan (2008 Stockholm Water Prize Laureate). The most elaborate publications on how to estimate water footprints are a 2004 report on the Water footprint of nations from UNESCO-IHE and the 2008 book *Globalization of Water*, and the 2011 manual "The water footprint assessment manual: Setting the global standard". Cooperation between global leading institutions in the field has led to the establishment of the Water Footprint Network in 2008 that aims to coordinate efforts to further develop and disseminate knowledge on water footprint concepts, methods and tools.

Water Footprint Network (WFN)

The Water Footprint Network is an international learning community (non-profit foundation under Dutch law) that serves as a platform for connecting communities interested in sustainability, equitability and efficiency of water use. The organization has two work programmes: a Technical Work Programme and a Policy Work Programme. In addition, there is a Partner Forum which offer partners of the WFN a way of receiving, contributing and exchanging knowledge and experience on water footprint. Its mission and activities are listed below and taken directly from the Water Footprint website.

Blue Water Footprint

The *blue water footprint* is the volume of freshwater that evaporated from the global blue water resources (surface water and ground water) to produce the goods and services consumed by the individual or community (either lost through evapotranspiration, incorporated in products or transferred to non-blue catchments).

Green Water Footprint

The *green water footprint* is the volume of water evaporated from the global green water resources (rainwater stored in the soil as soil moisture) during production or those incorporated in products.

Grey Water Footprint

The *grey water footprint* is the volume of polluted water that associates with the production of all goods and services for the individual or community. The latter can be estimated as the volume of water that is required to dilute pollutants to such an extent that the quality of the water remains at or above agreed water quality standards. It is calculated as:

$$WF_{proc,grey} = \frac{L}{c_{max} - c_{nat}}$$

where L is the pollutant load (as mass flux), c_{max} the maximum allowable concentration and c_{nat} the natural concentration of the pollutant in the receiving water body (both expressed in mass/volume).

Sustainable Water use

Sustainable water use involves the rigorous assessment of all source of clean water to establish the current and future rates of use, the impacts of that use both downstream and in the wider area where the water may be used and the impact of contaminated water streams on the environment and economic well being of the area. It also involves the implementation of social policies such as water pricing in order to manage water demand. In some localities, water may also have spiritual relevance and the use of such water may need to take account of such interests. For example, the Maori believe that water is the source and foundation of all life and have many spiritual associations with water and places associated with water. On a national and global scale, water sustainability requires strategic and long term planning to ensure appropriate sources of clean water are identified and the environmental and economic impact of such choices are understood and accepted. The re-use and reclamation of water is also part of sustainability including downstream impacts on both surface waters and ground waters.

CALCULATION FOR DIFFERENT ACTORS

The water footprint of a process is expressed as volumetric flow rate of water. That of a product is the whole footprint (sum) of processes in its complete supply chain divided by the number of product units. For consumers, businesses and geographic area, water footprint is indicated as volume of water per time, in particular:

- That of a consumer is the sum of footprint of all consumed products.
- That of a community or a nation is the sum for all of its members resp. inhabitants.
- That of a business is the footprint of all produced goods.
- That of a geographically delineated area is the footprint of all processes undertaken in this area. The virtual water balance of an area is the net import of virtual water $V_{i,\ net}$, defined as the difference of the gross import V_i of virtual water from its gross export V_e. The water footprint of national consumption $WF_{area,nat}$ results from this as the sum of the water footprint of national area and its virtual water balance.

International Standard

In February 2011, the Water Footprint Network, in a global collaborative effort of environmental organizations, companies, research institutions and the UN, launched the Global Water Footprint Standard. On July 24, 2014, ISO issued ISO 14046:2014, Environmental management—Water footprint—Principles, requirements and guidelines. The ISO standard is based on Life Cycle Assessment (LCA) principles and can be applied for different sorts of assessment of products and companies.

LIFE CYCLE ANALYSIS OF WATER USE

Life Cycle Analysis (LCA) is a systematic, phased approach to assessing the environmental aspects and potential impacts that are associated with a product, process or service.

"Life cycle" "refers to the major activities in the course of the product's life-span from its manufacture, use, and maintenance, to its final disposal, including the raw material acquisition required to manufacture the product". Thus a method for assessing the environmental impacts of freshwater consumption was developed. It specifically looks at the damage to three areas of protection: human health, ecosystem quality, and resources. The consideration of water consumption is crucial where water-intensive products (for example agricultural goods) are concerned and need to therefore undergo a life-cycle assessment. In addition, regional assessments are equally as necessary as the impact of water use depends on its location. In short, LCA is important as it identifies the impact of water use in certain products, consumers, companies, nations, etc. which can help reduce the amount of water used.

WATER FOOTPRINT OF PRODUCTS

The water footprint of a product is the total volume of freshwater used to produce the product, summed over the various steps of the production chain. The water footprint of a product refers not only to the total volume of water used; it also refers to where and when the water is used (Source: WFN Glossary). The Water Footprint Network maintains a global database on the water footprint of products: WaterStat

The water footprints involved in various diets vary greatly, and much of the variation tends to be associated with levels of meat consumption.The following table gives some examples of estimated global average water footprints of some agricultural products.

Product	Global average water footprint, L/kg
almonds, shelled	16,194
beef	15,415
chocolate	17,196
cotton lint	9,114
lettuce	238
milk	1,021
olive oil	14,430
tomatoes, fresh	214
tomatoes, dried	4,275
vanilla beans	126,505
wheat bread	1,608

(For more product water footprints: see the Product Gallery of the Water Footprint Network)

WATER FOOTPRINT OF INDIVIDUAL CONSUMERS

The water footprint of an individual refers to the sum of his or her direct and indirect freshwater use. The direct water use is the water used at home, while the indirect water use relates to the total volume of freshwater that is used to produce the goods and services consumed.

The average global water footprint of an individual is 1,385 m^3 per year. Residents in some example nations have a water footprints as shown in the table:

Nation	**annual water footprint**
China	1,071 m^3
Finland	1,733 m^3
India	1,089 m^3
United Kingdom	1,695 m^3
United States	2,842 m^3

WATER FOOTPRINT OF COMPANIES

The water footprint of a business, the 'corporate water footprint', is defined as the total volume of freshwater that is used directly or indirectly to run and support a business. It is the total volume of water use to be associated with the use of the business outputs. The water footprint of a business consists of water used for producing/manufacturing or for supporting activities and the indirect water use in the producer's supply chain.

The Carbon Trust argue that a more robust approach is for businesses to go beyond simple volumetric measurement to assess the full range of water impact from all sites. Its work with leading global pharmaceutical company GlaxoSmithKline (GSK) analysed four key categories: water availability, water quality, health impacts, and licence to operate (including reputational and regulatory risks) in order to enable GSK to quantitatively measure, and credibly reduce, its year-on-year water impact.

The Coca-Cola Company operates over a thousand manufacturing plants in about 200 countries. Making its drink uses a lot of water. Critics say its water footprint has been large. Coca-Cola has started to look at its water sustainability. It has now set out goals to reduce its water footprint such as treating the water it uses so it goes back into the environment in a clean state. Another goal is to find sustainable sources for the raw materials it uses in its drinks, such as sugarcane, oranges, and corn. By making its water footprint better, the company can reduce costs, improve the environment, and benefit the communities in which it operates.

WATER FOOTPRINT OF NATIONS

The water footprint of a nation is the water used to produce the goods and services consumed by the inhabitants of the nation. The internal water footprint is the appropriation of domestic water resources; the external water footprint is the appropriation of water resources in other countries. About 65% of Japan's total water footprint comes from outside the country; about 7% of the Chinese water footprint falls outside China.

Europe

Each EU citizen consumes 4,815 litres of water per day on average, 44% is used in power production primarily to cool thermal plants or nuclear power plants. Energy production annual water consumption in the EU 27 in 2011 was as billion m3 for gas 0.53, coal 1.54 and nuclear 2.44. Wind energy avoided the use of 387 million cubic metres (mn m^3) of water in 2012 avoiding a cost of €743 mn. The cost of droughts in Europe over the past thirty years is according to the European Commission €100 billion.

CRITICISM OF WATER FOOTPRINT AND VIRTUAL WATER

Insufficient Consideration of Consequences of Proposed Water Saving Policies to Farm Households

According to Dennis Wichelns of the International Water Management Institute: *Although one goal of virtual water analysis is to describe opportunities for improving water security, there is almost no mention of the potential impacts of the prescriptions arising from that analysis on farm households in industrialized or developing countries. It is essential to consider more carefully the inherent flaws in the virtual water and water footprint perspectives, particularly when seeking guidance regarding policy decisions.*

Insufficient Consideration of Regional Water Scarcity

The application and interpretation of water footprints may sometimes be used to promote industrial activities that lead to facile criticism of certain products. For example, the 140 litres required for coffee production for one cup might be of no harm to water resources if its cultivation occurs mainly in humid areas, but could be damaging in more arid regions. Other factors such as hydrology, climate, geology, topography, population and demographics should also be taken into account. Nevertheless, high water footprint calculations do suggest that environmental concern may be appropriate.

The use of the term "footprint" can also confuse people familiar with the notion of a carbon footprint, because the water footprint concept includes sums of water quantities without necessarily evaluating related impacts. This is in contrast to the carbon footprint, where carbon emissions are not simply summarized but normalized by CO_2 emissions, which are globally identical, to account for the environmental harm. The difference is due to the somewhat more complex nature of water; while involved in the global hydrological cycle, it is expressed in conditions both local and regional through various forms like river basins, watersheds, on down to groundwater (as part of larger aquifer systems).

The water footprint of a business, the `corporate water footprint`, is defined as the total volume of fresh water that is used directly or indirectly to run and support a business. It is the total volume of water use to be associated with the use of the business outputs. The water footprint of a business consists of water used for producing/manufacturing or for supporting activities and the indirect water use in the producer's supply chain. Water Credit for conserving water: Nagpur based innovator Shripad Vaidya`s idea of giving water credit`s, much like carbon credits, to those who save and conserve water is gaining ground. These water credits can be marketed or sold to those in need of surplus water for social,agricultural or industrial ventures.

ENVIRONMENTAL WATER USE

Although agriculture's water use includes provision of important terrestrial environmental values (as discussed in the "Water footprint of products" section above), and much "green water' is used in maintaining forests and wild lands, there is also direct environmental use (e.g. of surface water) that may be allocated by governments. For example, in California, where water use issues are sometimes severe because of drought, about 48 percent of "dedicated water use" in an average water year is for the environment (somewhat more than for agriculture). Such environmental water use is for keeping streams flowing, maintaining aquatic and riparian habitats, keeping wetlands wet, etc.

SECTORAL DISTRIBUTIONS OF WITHDRAWN WATER USE

Several nations estimate sectoral distribution of use of water withdrawn from surface and groundwater sources. For example, in Canada, in 2005, 42 billion cu. m of withdrawn water were used, of which about 38 billion cu. m was freshwater. Distribution of this use among sectors was: thermoelectric power generation 66.2%, manufacturing 13.6%, residential 9.0%, agriculture 4.7%, commercial and institutional 2.7%, water treatment and distribution systems 2.3%, mining 1.1%, and oil and gas extraction 0.5%. The 38 billion cu.m of freshwater withdrawn in that year can be compared with the nation's annual freshwater yield (estimated as streamflow) of 3,472 billion cu.m. Sectoral distribution is different in many respects in the US, where agriculture accounts for about 39% of fresh water withdrawals, thermoelectric power generation 38%, industrial 4%, residential 1%, and mining (including oil and gas) 1%.

Within the agricultural sector, withdrawn water use is for irrigation and for livestock. Whereas all irrigation in the US (including loss in conveyance of irrigation water) is estimated to account for about 38 percent of US withdrawn freshwater use, the irrigation water used for production of livestock feed and forage has been estimated to account for about 9 percent, and other withdrawn freshwater use for the livestock sector (for drinking, washdown of facilities, etc.) is estimated at about 0.7 percent. Because agriculture is a major user of withdrawn water, changes in the magnitude and efficiency of its water use are important. In the US, from 1980 (when agriculture's withdrawn water use peaked) to 2010, there was a 23 percent reduction in agriculture's use of withdrawn water, while US agricultural output increased by 49 percent over that period.

In the US, irrigation water application data are collected in the quintennial Farm and Ranch Irrigation Survey, conducted as part of the Census of Agriculture. Such data indicate great differences in irrigation water use within various agricultural sectors. For example, about 14 percent of corn-for-grain land and 11 percent of soybean land in the US are irrigated, compared with 66 percent of vegetable land, 79 percent of orchard land and 97 percent of rice land.

LAND USE

Habitat fragmented by numerous roads near the Indiana Dunes National Lakeshore.

Land use involves the management and modification of natural environment or wilderness into built environment such as settlements and semi-natural habitats such as arable fields, pastures, and managed woods. It also has been defined as the total of arrangements, activities, and inputs that people undertake in a certain land cover type."

Regulation

A land use map of Europe—major non-natural land uses include arable farmland (yellow) and pasture (light green).

Land use practices vary considerably across the world. The United Nations' Food and Agriculture Organization Water Development Division explains that "Land use concerns the products and/or benefits obtained from use of the land as well as the land management actions (activities) carried out by humans to produce those products and benefits." As of the early 1990s, about 13% of the Earth was considered arable land, with 26% in pasture, 32% forests and woodland, and 1.5% urban areas.

As Albert Guttenberg (1959) wrote many years ago, `Land use` is a key term in the language of city planning. Commonly, political jurisdictions will undertake land-use planning and regulate the use of land in an attempt to avoid land-use conflicts. Land use plans are implemented through land division and use ordinances and regulations, such as zoning regulations. Management consulting firms and non-governmental organizations will frequently seek to influence these regulations before they are codified.

United States

In colonial America, few regulations existed to control the use of land, due to the seemingly endless amounts of it. As society shifted from rural to urban, public land regulation became important, especially to city governments trying to control industry, commerce, and housing within their boundaries. The first zoning ordinance was passed in New York City in 1916, and, by the 1930s, most states had adopted zoning laws. In the 1970s, concerns about the environment and historic preservation led to further regulation.

Today, federal, state, and local governments regulate growth and development through statutory law. The majority of controls on land, however, stem from the actions of private developers and individuals. Three typical situations bringing such private entities into the court system are: suits brought by one neighbor against another; suits brought by a public official against a neighboring landowner on behalf of the public; and suits involving individuals who share ownership of a particular parcel of land. In these situations, judicial decisions and enforcement of private land-use arrangements can reinforce public regulation, and achieve forms and levels of control that regulatory zoning cannot.

Two major federal laws have been passed in the last half century that limit the use of land significantly. These are the National Historic Preservation Act of 1966 (today embodied in 16 U.S.C. 461 et seq.) and the National Environmental Policy Act of 1969 (42 U.S.C. 4321 et seq.).

Environment

Land use and land management practices have a major impact on natural resources including water, soil, nutrients, plants and animals. Land use information can be used to develop solutions for natural resource management issues such as salinity and water quality. For instance, water bodies in a region that has been deforested or having erosion will have different water quality than those in areas that are forested. Forest gardening, a plant-based food production system, is believed to be the oldest form of land use in the world.

The major effect of land use on land cover since 1750 has been deforestation of temperate regions. More recent significant effects of land use include urban sprawl, soil erosion, soil degradation, salinization, and desertification. Land-use change, together with use of fossil fuels, are the major anthropogenic sources of carbon dioxide, a dominant greenhouse gas.

According to a report by the United Nations` Food and Agriculture Organisation, land degradation has been exacerbated where there has been an absence of any land use planning, or of its orderly execution, or the existence of financial or legal incentives that have led to the wrong land use decisions, or one-sided central planning leading to over-utilization of the land resources - for instance for immediate production at all costs. As a consequence the result has often been misery for large segments of the local population and destruction of valuable ecosystems. Such narrow approaches should be replaced by a technique for the planning and management of land resources that is integrated and holistic and where land users are central. This will ensure the long-term quality of the land for human use, the prevention or resolution of social conflicts related to land use, and the conservation of ecosystems of high biodiversity value.

The citadel of Kastellet, Copenhagen that has been converted into a park, showing multiple examples of suburban land use

Urban Growth Boundaries

The urban growth boundary is one form of land-use regulation. For example, Portland, Oregon is required to have an urban growth boundary which contains at least 20,000 acres (81 km^2) of vacant land. Additionally, Oregon restricts the development of farmland. The regulations are controversial, but an economic analysis concluded that farmland appreciated similarly to the other land.

CULTURAL LANDSCAPE

Sunrise Matobo Hills, Zimbabwe.

A **cultural landscape**, as defined by the World Heritage Committee, is the "cultural properties [that] represent the combined works of nature and of man."

- "a landscape designed and created intentionally by man"
- an organically evolved landscape which may be a relict (or fossil) landscape" or a "continuing landscape"
- an associative cultural landscape which may be valued because of the religious, artistic or cultural associations of the natural element.

History of the Concept

The concept of 'cultural landscapes' can be found in the European tradition of landscape painting. From the 16th century onwards, many European artists painted landscapes in favor of people, diminishing the people in their paintings to figures subsumed within broader, regionally specific landscapes.

The word "landscape" itself combines 'land' with a verb of Germanic origin, "*scapjan/ schaffen*" to mean, literally, 'shaped lands'. Lands were then considered shaped by natural forces, and the unique details of such *landshaffen* (shaped lands) became themselves the subject of 'landscape' paintings.

The geographer Otto Schlüter is credited with having first formally used "cultural landscape" as an academic term in the early 20th century. In 1908, Schlüter argued that by defining geography as a *Landschaftskunde* (landscape science) this would give geography a logical subject matter shared by no other discipline.He defined two forms of landscape: the *Urlandschaft* (transl. original landscape) or landscape that existed before major human induced changes and the *Kulturlandschaft* (transl. 'cultural landscape') a landscape created by human culture. The major task of geography was to trace the changes in these two landscapes.

It was Carl O. Sauer, a human geographer, who was probably the most influential in promoting and developing the idea of cultural landscapes. Sauer was determined to stress the agency of culture as a force in shaping the visible features of the Earth's surface in delimited areas. Within his definition, the physical environment retains a central significance, as the medium with and through which human cultures act. His classic definition of a `cultural landscape` reads as follows:

"The cultural landscape is fashioned from a natural landscape by a cultural group. Culture is the agent, the natural area is the medium, the cultural landscape is the result

Since Schlüter's first formal use of the term, and Sauer's effective promotion of the idea, the concept of 'cultural landscapes has been variously used, applied, debated, developed and refined within academia, when, in 1992, the World Heritage Committee elected to convene a meeting of the 'specialists' to advise and assist redraft the Committee's Operational Guidelines to include 'cultural landscapes' as an option for heritage listing properties that were neither purely natural nor purely cultural in form (i.e. 'mixed' heritage).

The World Heritage Committee's adoption and use of the concept of 'cultural landscapes' has seen multiple specialists around the world, and many nations identifying 'cultural landscapes', assessing 'cultural landscapes', heritage listing 'cultural landscapes', managing 'cultural landscapes', and effectively making 'cultural landscapes' known and visible to the world, with very practical ramifications and challenges.

A 2006 academic review of the combined efforts of the World Heritage Committee, multiple specialists around the world, and nations to apply the concept of 'cultural landscapes', observed and concluded that:

"Although the concept of landscape has been unhooked for some time from its original art associations .. there is still a dominant view of landscapes as an inscribed surface, akin to a map or a text, from which cultural meaning and social forms can simply be read."

Within academia, any system of interaction between human activity and natural habitat is regarded as a cultural landscape. In a sense this understanding is broader than the definition applied within UNESCO, including, as it does, almost the whole of the world's occupied surface, plus almost all the uses, ecologies, interactions, practices, beliefs, concepts, and traditions of people living within cultural landscapes.

Some Universities now offer specialist degrees in the study of cultural landscapes, including, for instance, the Universities of Naples, St.-Étienne, and Stuttgart who offer a Master of Cultural Landscapes diploma.

Examples

The World Heritage Committee has identified and listed a number of areas or properties as cultural landscapes of universal value to humankind, including the following:

Tongariro National Park, New Zealand (1993)

Tongariro Mahuia River in the Tongariro National Park, New Zealand.

"In 1993 Tongariro National Park, became the first property to be inscribed on the World Heritage List under the revised criteria describing cultural landscapes. The mountains at the heart of the park have cultural and religious significance for the Maori people and symbolize the spiritual links between this community and its environment. The park has active and extinct volcanoes, a diverse range of ecosystems and some spectacular landscapes."

Uluru-Kata Tjuta National Park, Australia (1994)

"This park, formerly called Uluru (Ayers Rock – Mount Olga) National Park, features spectacular geological formations that dominate the vast red sandy plain of central

Australia. Uluru, an immense monolith, and Kata Tjuta, the rock domes located west of Uluru, form part of the traditional belief system of one of the oldest human societies in the world. The tradit0ional owners of Uluru-Kata Tjuta are the Anangu Aboriginal people."

Rice Terraces of Philippine Cordilleras (1995)

The Batad rice terraces, The Rice Terraces of the Philippine Cordilleras, the first site to be included in the UNESCO World Heritage List cultural landscape category in 1995.

For 2,000 years, the high rice fields of the Ifugao have followed the contours of the mountains. The fruit of knowledge handed down from one generation to the next, and the expression of sacred traditions and a delicate social balance, they have helped to create a landscape of great beauty that expresses the harmony between humankind and the environment.

Cultural Landscape of Sintra Portugal (1995)

"In the 19th century Sintra became the first centre of European Romantic architecture. Ferdinand II turned a ruined monastery into a castle where this new sensitivity was displayed in the use of Gothic, Egyptian, Moorish and Renaissance elements and in the creation of a park blending local and exotic species of trees. Other fine dwellings, built along the same lines in the surrounding serra, created a unique combination of parks and gardens which influenced the development of landscape architecture throughout Europe".

Portovenere, Cinque Terre, and the Islands (Palmaria, Tino and Tinetto), Italy (1997)

"The Ligurian coast between Cinque Terre and Portovenere is a cultural landscape of great scenic and cultural value. The layout and disposition of the small towns and the shaping of the surrounding landscape, overcoming the disadvantages of a steep, uneven terrain, encapsulate the continuous history of human settlement in this region over the past millennium."

Hortobágy National Park, Hungary (1999)

Hungarian Grey Cattle in Hortobágy National Park.

Hortobágy National Park is the largest continuous natural grassland in Europe, which means that it was not formed as a result of deforestation or river control. The first Hungarian national park (established in 1973), it is the country's largest protected area (82 thousand hectares). A significant part of it is Biosphere Reserve and a quarter of its area enjoys international protection under the Ramsar Convention on the conservation of wetlands.

Matopo Hills, Zimbabwe (2003)

The area exhibits a profusion of distinctive rock landforms rising above the granite shield that covers much of Zimbabwe. The large boulders provide abundant natural shelters and have been associated with human occupation from the early Stone Age right through to early historical times, and intermittently since. They also feature an outstanding collection of rock paintings. The Matopo Hills continue to provide a strong focus for the local community, which still uses shrines and sacred places closely linked to traditional, social and economic activities.

Dresden Elbe Valley, Germany (2004)

"The 18th- and 19th-century cultural landscape of Dresden Elbe Valley ... features low meadows, and is crowned by the Pillnitz Palace and the centre of Dresden with its numerous monuments and parks from the 16th to 20th centuries. The landscape also features 19th- and 20th-century suburban villas and gardens and valuable natural features."

This landscape was struck from the World Heritage list in 2009, due to the construction of a four lane highway across the Elbe.

Qhapaq Ñan (Andean Road System), Peru, Columbia, Equator, Bolivia, Argentina, Chile (2014)

This site is an extensive Inca communication, trade and defence network of roads covering 30,000 km. Constructed by the Incas over several centuries and partly based on pre-Inca infrastructure, this extraordinary network through one of the world's most extreme geographical terrains linked the snow-capped peaks of the Andes – at an altitude of more than 6,000 m – to the coast, running through hot rainforests, fertile valleys and absolute deserts. It reached its maximum expansion in the 15th century, when it spread across the length and breadth of the Andes. The Qhapac Ñan, Andean Road System includes 273 component sites spread over more than 6,000 km that were selected to highlight the social, political, architectural and engineering achievements of the network, along with its associated infrastructure for trade, accommodation and storage, as well as sites of religious significance.

The West Lake of HangzhouWorld Heritage Site. According to the UNESCO, it reflects an idealized fusion between humans and nature.

West Lake Cultural Landscape of Hangzhou, China (2011)

"The West Lake Cultural Landscape of Hangzhou, comprising the West Lake and the hills surrounding its three sides, has inspired famous poets, scholars and artists since the 9th century. It comprises numerous temples, pagodas, pavilions, gardens and ornamental trees, as well as causeways and artificial islands.

REFERENCES

1. Schneider, U. et al. 2014. GPCC's new land surface precipitation climatology based on quality-controlled in-situ data and its role in quantifying the global water cycle. Theoretical and Applied Climatology. 115: 15-40.
2. FAO. Water use. http://www.fao.org/nr/water/aquastat/water_use/index.stm#tables
3. Frenken, K. and V. Gillet. 2012. Irrigation water requirement and water withdrawal by country. AQUASTAT, FAO.
4. FAO. 2014. Water withdrawal by sector, around 2007. http://www.fao.org/nr/water/aquastat/tables/WorldData-Withdrawal_eng.pdf
5. "Looming water crisis simply a management problem" by Jonathan Chenoweth, New Scientist 28 Aug., 2008, pp. 28-32.
6. Mekonnen, M.M. and Hoekstra, A.Y.,National water footprint accounts: the green, blue and grey water footprint of production and consumption, Value of Water Research Report Series No. 50, UNESCO-IHE, Delft, the Netherlands. (PDF). Water Footprint Network. 2011. Archived from the original on 2014.
7. Cashing in on climate change . IBISWorld. 29 May 2008. Archived from the original on 4 October 2008.
8. Definition taken from the Hoekstra, A.Y. and Chapagain, A.K. (2008) Globalization of water: Sharing the planet`s freshwater resources, Blackwell Publishing, Oxford, UK.
9. Waterfootprint.org: Water footprint and virtual water . The Water Footprint Network. Retrieved 9 April 2014.
10. Hoekstra, A.Y. (2003) (ed) Virtual water trade: Proceedings of the International Expert Meeting on Virtual Water Trade, IHE Delft, the Netherlands
11. Globalization of Water, A.Y. Hoekstra and A.K. Chapagain, Blackwell, 2008
12. Hoekstra, Arjen (2011). The water footprint assessment manual: Setting the global standard (PDF). London: Earthscan. ISBN 978-1-84971-279-8.
13. Why a Water Footprint Network . Water Footprint Network. Retrieved 2013. Check date values in: |access-date= (help)
14. The Water Footprint Assessment Manual . Water Footprint Network. Retrieved 2015-01-20.
15. Policies and measure to promote sustainable water use . Europeanm Environment Agency. 18 February 2008. Retrieved 26 April 2016.
16. e Ahukaramū Charles Royal (22 September 2012). Tangaroa – the sea - Water as the source of life . Encyclopaedia of New Zealand.
17. Water Consumption and Sustainable Water Resources Management . OECD Library. 25 March 1998. ISBN 9789264162648. Retrieved 26 April 2016.
18. Naumann, Ruth (2011). Sustainability (1st ed.). North Shore, N.Z.: Cengage Learning. pp. 56–58. ISBN 978-017021-034-8.

19. Scientific Applications International Corporations (SAIC) (2006). Life Cycle Assessment: Principles and Practice. Reston, VA: SAIC.

20. Pfister, Stephan; Koehler, Annette; Hellweg, Stefanie (20 March 2009). Assessing the Environmental Impacts of Freshwater Consumption in LCA . Environmental Science. 43: 4008–104.

21. Vanham, D., M. M.Mekonnen and A. Y. Hoekstra. 2013. The water footprint of the EU for different diets. Ecological Indicators 32: 1-8.

22. Mekonnen, M. M. and A. Y. Hoekstra. 2010. The green, blue and grey water footprint of farm animals and animal products. Volume 1: Main report. UNESCO-IHE., Institute for Water Education. 50 pp.

23. Mekonnen, M. M. and A. Y. Hoekstra. 2010. The green, blue and grey water footprint of crops and derived crop products. Volume 2. Appendices main report. Value of Water Research Report Series No. 47. UNESCO-IHE Institute for Water Education. 1196 pp.

24. Hoekstra, AY (2012). The Water Footprint of Humanity . PNAS. doi:10.1073/pnas.1109936109.

25. Data obtained from the Finnish Wikipedia article page Vesijalanjälki

26. Chapagain, A.K. & Orr, S. U.K. Water Footprint: The Impact of the U.K.`s Food and Fibre Consumption on Global Water Resources, Volume 1 (PDF). WWF-UK. WWF-UK. and volume 2 Chapagain, A.K. & Orr, S. "Volume 2" (PDF). WWF-UK. WWF-UK.

27. The Water Footprint of Humanity . JournalistsResource.org, retrieved 20 March 2012

28. "Water, water everywhere... or is it? , The Carbon Trust, 26 November 2014. Retrieved on 20 January 2015.

29. 2013 Water Report: The Coca-Cola Company . The Coca-Cola Company. Retrieved 8 April 2014.

30. Saving water with wind energy EWEA June 2014

31. Saving water with wind energy Summary EWEA

32. Wichelns, Dennis (2010). Virtual water and water footprints offer limited insight regarding important policy questions . International Journal of Water Resources Development. 26 (4): 639–651. doi:10.1080/07900627.2010.519494 . Retrieved 2015-01-21.

33. http://www.scribd.com/doc/97302583/Water-use

34. Limca Book of Records2012 page 278

35. https://arnoneumann.wordpress.com/tag/environment-2/page/2/

36. California Department of Water Resources. California State Water Project water supply. http://www.water.ca.gov/swp/watersupply.cfm

37. Statistics Canada. 2010. Human activity and the environment. Freshwater supply and demand in Canada. Catalogue no. 16-201-X.

38. Maupin, M. A. et al. 2014. Estimated use of water in the United States 2010. U. S. Geological Survey Circular 1405. 55 pp.

39. Zering, K. D., T. J. Centner, D. Meyer, G. L. Newton, J. M. Sweeten and S. Woodruff. 2012. Water and land issues associated with animal agriculture: a U.S. perspective. CAST Issue Paper No. 50. Council for Agricultural Science and Technology, Ames, Iowa. 24 pp.

40. USA ERS.2013. Table 1. Indices of farm output, input and total factor productivity for the United States, 1948-2011. (last update 9/27/2013) http://www.ers.usda.gov/data-products/agricultural-productivity-in-the-us.aspx#28247

41. US Department of Agriculture. 2009. 2007 Census of agriculture. Farm and ranch irrigation survey (2008). Volume 3. Special Studies. Part 1. AC-07-SS-1. 177 pp. + appendices.

42. USDA. 2009. 2007 Census of agriculture. United States summary and State Data. Vol. 1. Geographic Area Series. Part 51. AC-07-A-51. 639 pp. + appendices.

CHAPTER 3

Irrigation Management

Chapter Contents

- **Introduction**
- **Water Sources**
- **Efficiency**
- **Technical Challenges**
- **Irrigation Sprinkler**
- **Gallery**
- **Environmental Impact of Irrigation**
- **Adverse Impacts**

Irrigation is the process by which water is brought to dry land through artificial means, such as pipes, hoses or ditches. The land that is being irrigated usually contains crops, grass or vegetation that would not usually receive enough water from rainfall or other natural sources. Sometimes the reason to irrigate a portion of land is that it happens to be a dry season, with less-than-average amounts of rainfall, or it might be necessary to do so because that land never would receive enough water on its own to be fertile. The water that is used for irrigation might be taken from nearby lakes, reservoirs, rivers or wells.

INTRODUCTION

Irrigation is the method in which a controlled amount of water is supplied to plants at regular intervals for agriculture. It is used to assist in the growing of agricultural crops, maintenance of landscapes, and revegetation of disturbed soils in dry areas and during periods of inadequate rainfall. Additionally, irrigation also has a few other uses in crop production, which include protecting plants against frost, suppressing weed growth in grain fields and preventing soil consolidation. In contrast, agriculture that relies only on direct rainfall is referred to as rain-fed or dry land farming.

An irrigation sprinkler watering a lawn.

Irrigation canal in Osmaniye, Turkey.

Irrigation systems are also used for dust suppression, disposal of sewage, and in mining. Irrigation is often studied together with drainage, which is the natural or artificial removal of surface and sub-surface water from a given area.

Irrigation has been a central feature of agriculture for over 5,000 years and is the product of many cultures. Historically, it was the basis for economies and societies across the globe, from Asia to the Southwestern United States.

History

Animal-powered irrigation, Upper Egypt, ca. 1846

Inside a karez tunnel at Turpan, Sinkiang.

Archaeological investigation has found evidence of irrigation where the natural rainfall was insufficient to support crops for rainfed agriculture.

Perennial irrigation was practiced in the Mesopotamian plain whereby crops were regularly watered throughout the growing season by coaxing water through a matrix of small channels formed in the field.

irrigation in Tamil Nadu (India).

Ancient Egyptians practiced *Basin irrigation* using the flooding of the Nile to inundate land plots which had been surrounded by dykes. The flood water was held until the fertile sediment had settled before the surplus was returned to the watercourse. There is evidence of the ancient Egyptian pharaoh Amenemhet III in the twelfth dynasty (about 1800 BCE) using the natural lake of the Faiyum Oasis as a reservoir to store surpluses of water for use during the dry seasons, the lake swelled annually from flooding of the Nile.

The Ancient Nubians developed a form of irrigation by using a waterwheel-like device called a *sakia*. Irrigation began in Nubia some time between the third and second millennium BCE. It largely depended upon the flood waters that would flow through the Nile River and other rivers in what is now the Sudan.

In sub-Saharan Africa irrigation reached the Niger River region cultures and civilizations by the first or second millennium BCE and was based on wet season flooding and water harvesting.

Terrace irrigation is evidenced in pre-Columbian America, early Syria, India, and China. In the Zana Valley of the Andes Mountains in Peru, archaeologists found remains of three irrigation canals radiocarbon dated from the 4th millennium BCE, the 3rd millennium BCE and the 9th century CE. These canals are the earliest record of irrigation in the New World. Traces of a canal possibly dating from the 5th millennium BCE were found under the 4th millennium canal. Sophisticated irrigation and storage systems were developed by the Indus Valley Civilization in present-day Pakistan and North India, including the reservoirs at Girnar in 3000 BCE and an early canal irrigation system from circa 2600 BCE. Large scale agriculture was practiced and an extensive network of canals was used for the purpose of irrigation.

Ancient Persia (modern day Iran) as far back as the 6th millennium BCE, where barley was grown in areas where the natural rainfall was insufficient to support such a crop. The Qanats, developed in ancient Persia in about 800 BCE, are among the oldest known irrigation methods still in use today. They are now found in Asia, the Middle East and North Africa. The system comprises a network of vertical wells and gently sloping tunnels driven into the sides of cliffs and steep hills to tap groundwater. The noria, a water wheel with clay pots around the rim powered by the flow of the stream (or by animals where the water source was still), was first brought into use at about this time, by Roman settlers in North Africa. By 150 BCE the pots were fitted with valves to allow smoother filling as they were forced into the water.

The irrigation works of ancient Sri Lanka, the earliest dating from about 300 BCE, in the reign of King Pandukabhaya and under continuous development for the next thousand years, were one of the most complex irrigation systems of the ancient world. In addition to underground canals, the Sinhalese were the first to build completely artificial reservoirs to store water. Due to their engineering superiority in this sector, they were often called 'masters of irrigation'. Most of these irrigation systems still exist undamaged up to now, in Anuradhapura and Polonnaruwa, because of the advanced and precise engineering. The system was extensively restored and further extended during the reign of King Parakrama Bahu (1153–1186 CE).

China

The oldest known hydraulic engineers of China were Sunshu Ao (6th century BCE) of the Spring and Autumn period and Ximen Bao (5th century BCE) of the Warring States period, both of whom worked on large irrigation projects. In the Sichuan region belonging to the State of Qin of ancient China, the Dujiangyan Irrigation System was built in 256 BCE to irrigate an enormous area of farmland that today still supplies water. By the 2nd century AD, during the Han Dynasty, the Chinese also used chain pumps that lifted water from lower elevation to higher elevation. These were powered by manual foot pedal, hydraulic waterwheels, or rotating mechanical wheels pulled by oxen. The water was used for public works of providing water for urban residential quarters and palace gardens, but mostly for irrigation of farmland canals and channels in the fields.

Korea

In 15th century Korea, the world's first rain gauge, *uryanggye* (Korean:우량계), was invented in 1441. The inventor was Jang Yeong-sil, a Korean engineer of the Joseon Dynasty, under the active direction of the king, Sejong the Great. It was installed in irrigation tanks as part of a nationwide system to measure and collect rainfall for agricultural applications. With this instrument, planners and farmers could make better use of the information gathered in the survey.

North America

In North America, the Hohokam were the only culture known to rely on irrigation canals to water their crops, and their irrigation systems supported the largest population in the Southwest by AD 1300. The Hohokam constructed an assortment of simple canals combined with weirs in their various agricultural pursuits. Between the 7th and 14th centuries, they also built and maintained extensive irrigation networks along the lower Salt and middle Gila rivers that rivaled the complexity of those used in the ancient Near East, Egypt, and China. These were constructed using relatively simple excavation tools, without the benefit of advanced engineering technologies, and achieved drops of a few feet per mile, balancing erosion and siltation. The Hohokam cultivated varieties of cotton, tobacco, maize, beans and squash, as well as harvested an assortment of wild plants. Late in the Hohokam Chronological Sequence, they also used extensive dry-farming systems, primarily to grow agave for food and fiber. Their reliance on agricultural strategies based on canal irrigation, vital in their less than hospitable desert environment and arid climate, provided the basis for the aggregation of rural populations into stable urban centers.

Present Extent

Irrigation ditch in Montour County, Pennsylvania, off Strawberry Ridge Road.

In the mid-20th century, the advent of diesel and electric motors led to systems that could pump groundwater out of major aquifers faster than drainage basins could refill them. This can lead to permanent loss of aquifer capacity, decreased water quality, ground subsidence, and other problems. The future of food production in such areas as the North China Plain, the Punjab, and the Great Plains of the US is threatened by this phenomenon.

At the global scale, 2,788,000 km^2 (689 million acres) of fertile land was equipped with irrigation infrastructure around the year 2000. About 68% of the area equipped for irrigation is located in Asia, 17% in the America, 9% in Europe, 5% in Africa and 1% in Oceania. The largest contiguous areas of high irrigation density are found:

- In Northern India and Pakistan along the Ganges and Indus rivers
- In the Hai He, Huang He and Yangtze basins in China
- Along the Nile river in Egypt and Sudan
- In the Mississippi-Missouri river basin and in parts of California

Smaller irrigation areas are spread across almost all populated parts of the world.

Only eight years later in 2008, the scale of irrigated land increased to an estimated total of 3,245,566 km^2 (802 million acres), which is nearly the size of India.

Types of Irrigation

Basin flood irrigation of wheat.

Irrigation of land in Punjab, Pakistan.

Various types of irrigation techniques differ in how the water obtained from the source is distributed within the field. In general, the goal is to supply the entire field uniformly with water, so that each plant has the amount of water it needs, neither too much nor too little.

Surface Irrigation

In *surface* (*furrow*, *flood*, or *level basin*) irrigation systems, water moves across the surface of agricultural lands, in order to wet it and infiltrate into the soil. Surface irrigation can be subdivided into furrow, *borderstrip or basin irrigation*. It is often called *flood irrigation* when the irrigation results in flooding or near flooding of the cultivated land. Historically, this has been the most common method of irrigating agricultural land and still used in most parts of the world.

Where water levels from the irrigation source permit, the levels are controlled by dikes, usually plugged by soil. This is often seen in terraced rice fields (rice paddies), where the method is used to flood or control the level of water in each distinct field. In some cases, the water is pumped, or lifted by human or animal power to the level of the land. The field water efficiency of surface irrigation is typically lower than other forms of irrigation but has the potential for efficiencies in the range of 70% - 90% under appropriate management.

Localized Irrigation

Impact sprinkler head.

Localized irrigation is a system where water is distributed under low pressure through a piped network, in a pre-determined pattern, and applied as a small discharge to each plant or adjacent to it. Drip irrigation, spray or micro-sprinkler irrigation and bubbler irrigation belong to this category of irrigation methods.

Subsurface Textile Irrigation

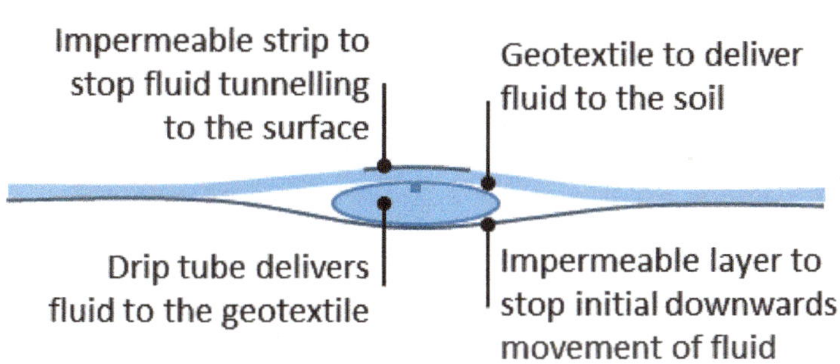

Diagram showing the structure of an example SSTI installation

Subsurface Textile Irrigation (SSTI) is a technology designed specifically for subsurface irrigation in all soil textures from desert sands to heavy clays. A typical subsurface textile irrigation system has an impermeable base layer (usually polyethylene or polypropylene), a drip line running along that base, a layer of geotextile on top of the drip line and, finally, a narrow impermeable layer on top of the geotextile (see diagram). Unlike standard drip irrigation, the spacing of emitters in the drip pipe is not critical as the geotextile moves the water along the fabric up to 2 m from the dripper.

Drip Irrigation

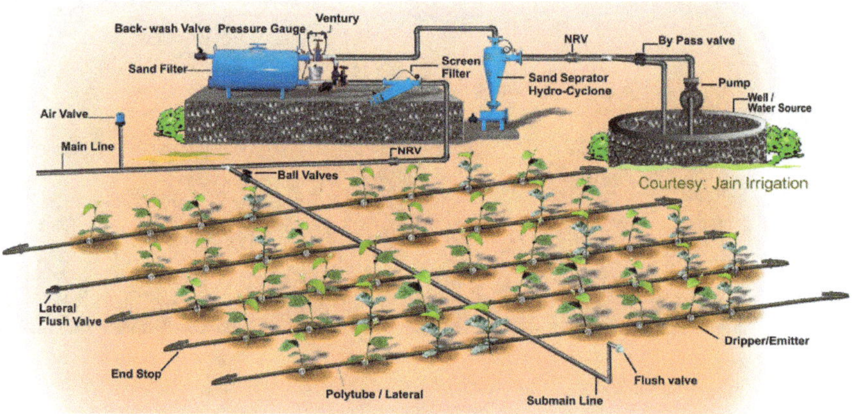

Drip irrigation layout and its parts.

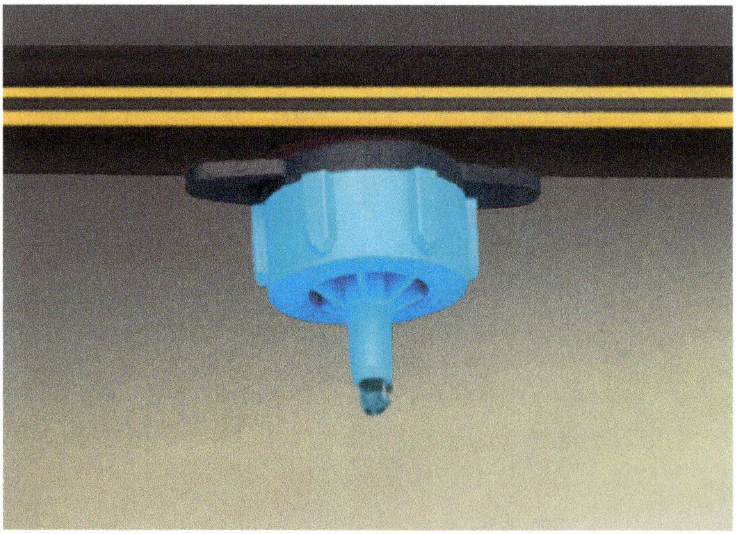

Drip irrigation – a dripper in action.

Grapes in Petrolina, only made possible in this semi arid area by drip irrigation.

Drip (or micro) irrigation, also known as trickle irrigation, functions as its name suggests. In this system water falls drop by drop just at the position of roots. Water is delivered at or near the root zone of plants, drop by drop. This method can be the most water-efficient method of irrigation, if managed properly, since evaporation and runoff are minimized. The field water efficiency of drip irrigation is typically in the range of 80 to 90 percent when managed correctly.

In modern agriculture, drip irrigation is often combined with plastic mulch, further reducing evaporation, and is also the means of delivery of fertilizer. The process is known as *fertigation*.

Deep percolation, where water moves below the root zone, can occur if a drip system is operated for too long or if the delivery rate is too high. Drip irrigation methods range from very high-tech and computerized to low-tech and labor-intensive. Lower water pressures are usually needed than for most other types of systems, with the exception of low energy center pivot systems and surface irrigation systems, and the system can be designed for uniformity throughout a field or for precise water delivery to individual plants in a landscape containing a mix of plant species. Although it is difficult to regulate pressure on steep slopes, pressure compensating emitters are available, so the field does not have to be level. High-tech solutions involve precisely calibrated emitters located along lines of tubing that extend from a computerized set of valves.

Irrigation using Sprinkler Systems

Sprinkler irrigation of blueberries in Plainville, New York, United States.

A traveling sprinkler at Millets Farm Centre, Oxfordshire, United Kingdom.

In *sprinkler* or overhead irrigation, water is piped to one or more central locations within the field and distributed by overhead high-pressure sprinklers or guns. A system utilizing sprinklers, sprays, or guns mounted overhead on permanently installed risers is often referred to as a *solid-set* irrigation system. Higher pressure sprinklers that rotate are called *rotors* an are driven by a ball drive, gear drive, or impact mechanism. Rotors can be designed to rotate in a full or partial circle. Guns are similar to rotors, except that they generally operate at very high pressures of 40 to 130 lbf/in^2 (275 to 900 kPa) and flows of 50 to 1200 US gal/min (3 to 76 L/s), usually with nozzle diameters in the range of 0.5 to 1.9 inches (10 to 50 mm). Guns are used not only for irrigation, but also for industrial applications such as dust suppression and logging.

Sprinklers can also be mounted on moving platforms connected to the water source by a hose. Automatically moving wheeled systems known as *traveling sprinklers* may irrigate areas such as small farms, sports fields, parks, pastures, and cemeteries unattended. Most of these utilize a length of polyethylene tubing wound on a steel drum. As the tubing is wound on the drum powered by the irrigation water or a small gas engine, the sprinkler is pulled across the field. When the sprinkler arrives back at the reel the system shuts off. This type of system is known to most people as a waterreel" traveling irrigation sprinkler and they are used extensively for dust suppression, irrigation, and land application of waste water.

Other travelers use a flat rubber hose that is dragged along behind while the sprinkler platform is pulled by a cable. These cable-type travelers are definitely old technology and their use is limited in today`s modern irrigation projects.

Irrigation using Center Pivot

A small center pivot system from beginning to end.

The hub of a center-pivot irrigation system

Rotator style pivot applicator sprinkler.

Center pivot with drop sprinklers

Wheel line irrigation system in Idaho, 2001.

Center pivot irrigation.

Center pivot irrigation is a form of sprinkler irrigation consisting of several segments of pipe (usually galvanized steel or aluminium) joined together and supported by trusses, mounted on wheeled towers with sprinklers positioned along its length. The system moves in a circular pattern and is fed with water from the pivot point at the center of the arc. These systems are found and used in all parts of the world and allow irrigation of all types of terrain. Newer systems have drop sprinkler heads as shown in the image that follows.

Most center pivot systems now have drops hanging from a u-shaped pipe attached at the top of the pipe with sprinkler head that are positioned a few feet (at most) above the crop, thus limiting evaporative losses. Drops can also be used with drag hoses or bubblers that deposit the water directly on the ground between crops. Crops are often planted in a circle to conform to the center pivot. This type of system is known as LEPA (Low Energy Precision Application). Originally, most center pivots were water powered. These were replaced by hydraulic systems (*T-L Irrigation*) and electric motor driven systems (Reinke, Valley, Zimmatic). Many modern pivots feature GPS devices.

Irrigation by Lateral move (side roll, wheel line, Wheelmove)

A *series of pipes, each with a wheel* of about 1.5 m diameter permanently affixed to its midpoint, and sprinklers along its length, are coupled together. Water is supplied at one end using a large hose. After sufficient irrigation has been applied to one strip of the field, the hose is removed, the water drained from the system, and the assembly rolled either by hand or with a purpose-built mechanism, so that the sprinklers are moved to a different position across the field. The hose is reconnected. The process is repeated in a pattern until the whole field has been irrigated.

This system is less expensive to install than a center pivot, but much more labor-intensive to operate - it does not travel automatically across the field: it applies water in a stationary strip, must be drained, and then rolled to a new strip. Most systems use 4 or 5-inch (130 mm) diameter aluminum pipe. The pipe doubles both as water transport and as an axle for rotating all the wheels. A drive system (often found near the centre of the wheel line) rotates the clamped-together pipe sections as a single axle, rolling the whole wheel line. Manual adjustment of individual wheel positions may be necessary if the system becomes misaligned.

Wheel line systems are limited in the amount of water they can carry, and limited in the height of crops that can be irrigated. One useful feature of a lateral move system is that it consists of sections that can be easily disconnected, adapting to field shape as the line is moved. They are most often used for small, rectilinear, or oddly-shaped fields, hilly or mountainous regions, or in regions where labor is inexpensive.

Sub-Irrigation

Subirrigation has been used for many years in field crops in areas with high water tables. It is a method of artificially raising the water table to allow the soil to be moistened from below the plants' root zone. Often those systems are located on permanent grasslands in lowlands or river valleys and combined with drainage infrastructure. A system of pumping stations, canals, weirs and gates allows it to increase or decrease the water level in a network of ditches and thereby control the water table.

Sub-irrigation is also used in commercial greenhouse production, usually for potted plants. Water is delivered from below, absorbed upwards, and the excess collected for recycling. Typically, a solution of water and nutrients floods a container or flows through a trough for a short period of time, 10–20 minutes, and is then pumped back into a holding tank for reuse. Sub-irrigation in greenhouses requires fairly sophisticated, expensive equipment and management. Advantages are water and nutrient conservation, and labor-saving through lowered system maintenance and automation. It is similar in principle and action to subsurface basin irrigation.

Irrigation Automatically, Non-Electric using Buckets and Ropes

Besides the common manual watering by bucket, an automated, natural version of this also exists. Using plain polyester ropes combined with a prepared ground mixture can be used to water plants from a vessel filled with water.

The ground mixture would need to be made depending on the plant itself, yet would mostly consist of black potting soil, vermiculite and perlite. This system would (with certain crops) allow to save expenses as it does not consume any electricity and only little water (unlike sprinklers, water timers, etc.). However, it may only be used with certain crops (probably mostly larger crops that do not need a humid environment; perhaps e.g. paprikas).

Irrigation using Water Condensed from Humid Air

In countries where at night, humid air sweeps the countryside.Water can be obtained from the humid air by condensation onto cold surfaces. This is for example practiced in the vineyards at Lanzarote using stones to condense water or with various fog collectors based on canvas or foil sheets.

In-ground Irrigation

Most commercial and residential irrigation systems are "in ground" systems, which means that everything is buried in the ground. With the pipes, sprinklers, emitters (drippers), and irrigation valves being hidden, it makes for a cleaner, more presentable landscape without garden hoses or other items having to be moved around manually. This does, however, create some drawbacks in the maintenance of a completely buried system.

Most irrigation systems are divided into zones. A zone is a single irrigation valve and one or a group of drippers or sprinklers that are connected by pipes or tubes. Irrigation systems are divided into zones because there is usually not enough pressure and available flow to run sprinklers for an entire yard or sports field at once. Each zone has a solenoid valve on it that is controlled via wire by an irrigation controller. The irrigation controller is either a mechanical (now the "dinosaur" type) or electrical device that signals a zone to turn on at a specific time and keeps it on for a specified amount of time. "Smart Controller" is a recent term for a controller that is capable of adjusting the watering time by itself in response to current environmental conditions. The smart controller determines current conditions by means of historic weather data for the local area, a soil moisture sensor (water potential or water content), rain sensor, or in more sophisticated systems satellite feed weather station, or a combination of these.

When a zone comes on, the water flows through the lateral lines and ultimately ends up at the irrigation emitter (drip) or sprinkler heads. Many sprinklers have pipe thread inlets on the bottom of them which allows a fitting and the pipe to be attached to them. The sprinklers are usually installed with the top of the head flush with the ground surface. When the water is pressurized, the head will pop up out of the ground and water the desired area until the valve closes and shuts off that zone. Once there is no more water pressure in the lateral line, the sprinkler head will retract back into the ground. Emitters are generally laid on the soil surface or buried a few inches to reduce evaporation losses.

WATER SOURCES

Irrigation water can come from groundwater (extracted from springs or by using wells), from surface water (withdrawn from rivers, lakes or reservoirs) or from non-conventional sources like treated wastewater, desalinated water or drainage water. A special form of irrigation using surface water is spate irrigation, also called floodwater harvesting. In case of a flood (spate), water is diverted to normally dry river beds (wadis) using a network of dams, gates and channels and spread over large areas. The moisture stored in the soil will be used thereafter to grow crops. Spate irrigation areas are in particular located in semi-arid or arid, mountainous regions. While floodwater harvesting belongs to the accepted irrigation methods, rainwater harvesting is usually not considered as a form of irrigation. Rainwater harvesting is the collection of runoff water from roofs or unused land and the concentration of this.

Irrigation is underway by pump-enabled extraction directly from the Gumti, seen in the background, in Comilla, Bangladesh.

Around 90% of wastewater produced globally remains untreated, causing widespread water pollution, especially in low-income countries. Increasingly, agriculture uses untreated wastewater as a source of irrigation water. Cities provide lucrative markets for fresh produce, so are attractive to farmers. However, because agriculture has to compete for increasingly scarce water resources with industry and municipal users (see Water scarcity below), there is often no alternative for farmers but to use water polluted with urban waste, including sewage, directly to water their crops. Significant health hazards can result from using water loaded with pathogens in this way, especially if people eat raw vegetables that have been irrigated with the polluted water. The International Water Management Institute has worked in India, Pakistan, Vietnam, Ghana, Ethiopia, Mexico and other countries on various projects aimed at assessing and reducing risks of wastewater irrigation. They advocate a 'multiple-barrier' approach to wastewater use, where farmers are encouraged to adopt various risk-reducing behaviours. These include ceasing irrigation a few days before harvesting to allow pathogens to die off in the sunlight, applying water carefully so it does not contaminate leaves likely to be eaten raw, cleaning vegetables with disinfectant or allowing fecal sludge used in farming to dry before being used as a human manure. The World Health Organization has developed guidelines for safe water use.

There are numerous benefits of using recycled water for irrigation, including the low cost (when compared to other sources, particularly in an urban area), consistency of supply (regardless of season, climatic conditions and associated water restrictions), and general consistency of quality. Irrigation of recycled wastewater is also considered as a means for plant fertilization and particularly nutrient supplementation. This approach carries with it a risk of soil and water pollution through excessive wastewater application. Hence, a detailed understanding of soil water conditions is essential for effective utilization of wastewater for irrigation.

EFFICIENCY

Young engineers restoring and developing the old Mughal irrigation system during the reign of the Mughal Emperor Bahadur Shah II.

Modern irrigation methods are efficient enough to supply the entire field uniformly with water, so that each plant has the amount of water it needs, neither too much nor too little. Water use efficiency in the field can be determined as follows:

- Field Water Efficiency (%) = (Water Transpired by Crop ÷ Water Applied to Field) x 100

Until 1960s, the common perception was that water was an infinite resource. At that time, there were fewer than half the current number of people on the planet. People were not as wealthy as today, consumed fewer calories and ate less meat, so less water was needed to produce their food. They required a third of the volume of water we presently take from rivers. Today, the competition for water resources is much more intense. This is because there are now more than seven billion people on the planet, their consumption of water-thirsty meat and vegetables is rising, and there is increasing competition for water from industry, urbanisation and biofuel crops. To avoid a global water crisis, farmers will have to strive to increase productivity to meet growing demands for food, while industry and cities find ways to use water more efficiently.

Successful agriculture is dependent upon farmers having sufficient access to water. However, water scarcity is already a critical constraint to farming in many parts of the world. With regards to agriculture, the World Bank targets food production and water management as an increasingly global issue that is fostering a growing debate. Physical water scarcity is where there is not enough water to meet all demands, including that needed for ecosystems to function effectively. Arid regions frequently suffer from physical water scarcity. It also occurs where water seems abundant but where resources are over-committed. This can happen where there is overdevelopment of hydraulic infrastructure, usually for irrigation. Symptoms of physical water scarcity include environmental degradation and declining groundwater. Economic scarcity, meanwhile, is caused by a lack of investment in water or insufficient human capacity to satisfy the demand for water. Symptoms of economic water scarcity include a lack of infrastructure, with people often having to fetch water from rivers for domestic and agricultural uses. Some 2.8 billion people currently live in water-scarce areas.

TECHNICAL CHALLENGES

Irrigation schemes involve solving numerous engineering and economic problems while minimizing negative environmental impact.

- Competition for surface water rights.
- Overdrafting (depletion) of underground aquifers.
- Ground subsidence (e.g. New Orleans, Louisiana)
- Underirrigation or irrigation giving only just enough water for the plant (e.g. in drip line irrigation) gives poor soil salinity control which leads to increased soil salinity with consequent buildup of toxic salts on soil surface in areas with high evaporation. This requires either leaching to remove these salts and a method of drainage to carry the salts away. When using drip lines, the leaching is best done regularly at certain intervals (with only a slight excess of water), so that the salt is flushed back under the plant's roots.
- Overirrigation because of poor distribution uniformity or management wastes water, chemicals, and may lead to water pollution.
- Deep drainage (from over-irrigation) may result in rising water tables which in some instances will lead to problems of irrigation salinity requiring watertable control by some form of subsurface land drainage.
- Irrigation with saline or high-sodium water may damage soil structure owing to the formation of alkaline soil
- Clogging of filters: It is mostly algae that clog filters, drip installations and nozzles. UV and ultrasonic method can be used for algae control in irrigation systems.

IRRIGATION SPRINKLER

Irrigation sprinklers are sprinklers providing irrigation to agriculture, crops, vegetation, or for recreation, as a cooling system, or for the control of airborne dust, landscaping and golf courses. The sprinkler system irrigates the field and thus it is widely used in sandy areas as it checks the wastage of water through seepage and evaporation. Sprinkler irrigation is a method of applying irrigation water which is similar to natural rainfall.

Water is distributed through a system of pipes usually by pumping. It is then sprayed into the air through sprinklers so that it breaks up into small water drops which fall to the ground. The pump supply system, sprinklers and operating conditions must be designed to enable a uniform application of water.

impact sprinkler head in action.

Types

Industrial

Sprinklers that spray in a fixed pattern are generally called sprays or spray heads. Sprays are not usually designed to operate at pressures , due to misting problems that may develop.

Higher pressure sprinklers that themselves move in a circle are driven by a ball drive, gear drive, or impact mechanism (impact sprinklers). These can be designed to rotate in a full or partial circle.

Rainguns are similar to impact sprinkler, except that they generally operate at very high pressures of 40 to 130 lbf/in^2 (275 to 900 kPa) and flows of 50 to 1200 US gal/min (3 to 76 L/s), usually with nozzle diameters in the range of 0.5 to 1.9 inches (10 to 50 mm). In addition to irrigation, guns are used for industrial applications such as dust suppression and logging.

Many irrigation sprinklers are buried in the ground along with their supporting plumbing, although above ground and moving sprinklers are also common. Most irrigation sprinklers operate through electric and hydraulic technology and are grouped together in zones that can be collectively turned on and off by actuating a solenoid-controlled valve.

Residential

An oscillating sprinkler is commonly used to water residential lawns, and is moved as needed.

Home lawn sprinklers vary widely in their size, cost, and complexity. They include impact sprinklers, oscillating sprinklers, drip sprinklers, and underground sprinkler systems. Small sprinklers are available at home and garden stores or hardware stores for small costs. These are often attached to an outdoor water faucet and are placed only temporarily. Other systems may be professionally installed permanently in the ground and are attached permanently to a home`s plumbing system. An ingenious domestic sprinkler made by Nomad called a 'set-and-forget tractor sprinkler' was used in Australia in the 1950s. Water pressure ensured that the sprinkler slowly moved across a lawn.

Permanently installed systems may often operate on timers or other automated processes. They are occasionally installed with retractable heads for aesthetic and practical reasons (making damage during lawn mowing or other maintenance less likely). These often are programmed to operate on a schedule and irrigate a piece of land in zones.

Underground Sprinkler

Underground sprinklers function through means of basic electronic and hydraulic technology. This valve and all of the sprinklers that will be activated by this valve are known as a zone. Upon activation, the solenoid, which sits on top of the valve is magnetized lifting a small stainless steel plunger in its center. By doing this, the activated (or raised) plunger allows air to escape from the top of a rubber diaphragm located in the center of the valve. Water that has been charged and waiting on the bottom of this same diaphragm now has the higher pressure and lifts the diaphragm. This pressurized water is then allowed to escape down stream of the valve through a series of pipes, usually made of PVC (higher pressure commercial systems) or polyethylene pipe (for typically lower pressure residential systems). At the end of these pipes and flush to ground level (typically) are pre measured and spaced out sprinklers. These sprinklers can be fixed spray heads that have a set pattern and generally spray between 1.5–2m (7–15 ft.), full rotating sprinklers that can spray a broken stream of water from 6–12m (20–40 ft.), or small drip emitters that release a slow, steady drip of water on more delicate plants such as flowers and shrubs. use of indigenous materials also recommended.

Agricultural Science

Center pivot irrigation **sprinkler nozzles, used in** crop irrigation

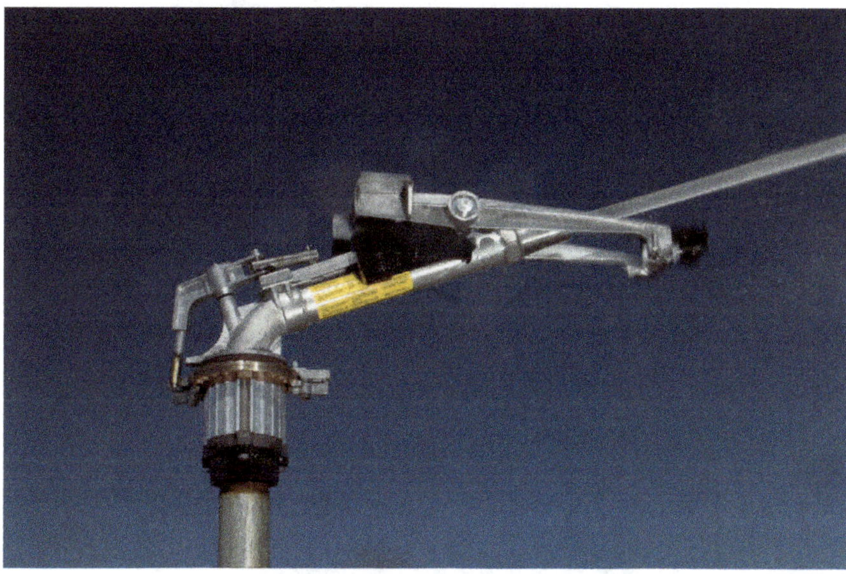

End Gun style pivot applicator sprinkler.

The first use of sprinklers by farmers was some form of home and golf course type sprinklers. These ad hoc systems, while doing the job of the buried pipes and fixed sprinkler heads, interfered with cultivation and were expensive to maintain. In the 1950s a firm based in Portland, Oregon *Stout-Wyss Irrigation System*, developed the rolling pipe type irrigation system for farms that has become the most popular type for farmers irrigating large fields. With this system large wheels attached to the large pipes with sprinkler heads move slowly across the field.

Use

Most irrigation sprinklers are used as part of a sprinkler system, consisting of various plumbing parts, pump unit, piping and control equipment. Outdoor sprinkler systems are sometimes used as a deterrent against homeless people. This sprinkler system was programmed to drench unsuspecting sleepers at random times during the night. Local businessmen soon copied this system in an effort to drive homeless people away from public sidewalks adjacent to their businesses.

GALLERY

An underground sprinkler system about to be installed.

Irrigation sprinkler watering crops.

Irrigation sprinkler.

A overhead sprinkler.

Sprinkler for children.

ENVIRONMENTAL IMPACT OF IRRIGATION

The **environmental impacts of irrigation** relate to the changes in quantity and quality of soil and water as a result of irrigation and the effects on natural and social conditions in river basins and downstream of an irrigation scheme. The impacts stem from the altered hydrological conditions caused by the installation and operation of the irrigation scheme.

Direct Effects

An irrigation scheme draws water from groundwater, rivers, lakes or overland flow, and distributes it over an area. Hydrological, or direct, effects of doing thisinclude reduction in downstream river flow, increased evaporation in the irrigated area, increased level in the water table as groundwater recharge in the area is increased and flow increased in the irrigated area. Likewise, irrigation has immediate effects on the provision of moisture to the atmosphere, inducing atmospheric instabilities and increasing downwind rainfall, or in other cases modifies the atmospheric circulation, delivering rain to different downwind areas. Increases or decreases in irrigation are a key area of concern in precipitationshed studies, that examine how significant modifications to the delivery of evaporation to the atmosphere can alter downwind rainfall.

Indirect Effects

Indirect effects are those that have consequences that take longer to develop and may also bė longer-lasting. The indirect effects of irrigation include the following:

- Waterlogging
- Soil salination
- Ecological damage
- Socioeconomic impacts

The indirect effects of waterlogging and soil salination occur directly on the land being irrigated. The ecological and socioeconomic consequences take longer to happen but can be more far-reaching.

Some irrigation schemes use water wells for irrigation. As a result, the overall water level decreases. This may cause water mining, land/soil subsidence, and, along the coast, saltwater intrusion.

Irrigated land area worldwide occupies about 16% of the total agricultural area and the crop yield of irrigated land is roughly 40% of the total yield. In other words, irrigated land produces 2.5 times more product than non-irrigated land. This article will discuss some of the environmental and socioeconomic impacts of irrigation.

ADVERSE IMPACTS

Reduced River Flow

The reduced downstream river flow may cause:

- reduced downstream flooding
- disappearance of ecologically and economically important wetlands or flood forests
- reduced availability of industrial, municipal, household, and drinking water
- reduced shipping routes. Water withdrawal poses a serious threat to the Ganges. In India, barrages control all of the tributaries to the Ganges and divert roughly 60 percent of river flow to irrigation
- reduced fishing opportunities. The Indus River in Pakistan faces scarcity due to

over-extraction of water for agriculture. The Indus is inhabited by 25 amphibian species and 147 fish species of which 22 are found nowhere else in the world. It harbors the endangered Indus River dolphin, one of the world's rarest mammals. Fish populations, the main source of protein and overall life support systems for many communities, are also being threatened

- reduced discharge into the sea, which may have various consequences like coastal erosion (e.g. in Ghana) and salt water intrusion in delta`s and estuaries (e.g. in Egypt, see Aswan dam). Current water withdrawal from the river Nile for irrigation is so high that, despite its size, in dry periods the river does not reach the sea. The Aral Sea has suffered an "environmental catastrophe" due to the interception of river water for irrigation purposes.

Increased Groundwater Recharge, Waterlogging, Soil Salinity

Looking over the shoulder of a Peruvian farmer in the Huarmey delta at waterlogged and salinised irrigated land with a poor crop stand.

This illustrates an environmental impact of upstream irrigation developments causing an increased flow of groundwater to this lower-lying area, leading to adverse conditions.

Increased groundwater recharge stems from the unavoidable deep percolation losses occurring in the irrigation scheme. The lower the irrigation efficiency, the higher the losses. Although fairly high irrigation efficiencies of 70% or more (i.e. losses of 30% or less) can occur with sophisticated techniques like sprinkler irrigation and drip irrigation, or by well managed surface irrigation, in practice the losses are commonly in the order of 40% to 60%. This may cause the following issues:

- rising water tables
- increased storage of groundwater that may be used for irrigation, municipal, household and drinking water by pumping from wells

- waterlogging and drainage problems in villages, agricultural lands, and along roads - with mostly negative consequences. The increased level of the water table can lead to reduced agricultural production.
- shallow water tables - a sign that the aquifer is unable to cope with the groundwater recharge stemming from the deep percolation losses
- where water tables are shallow, the irrigation applications are reduced. As a result, the soil is no longer leached and soil salinity problems develop
- stagnant water tables at the soil surface are known to increase the incidence of water-borne diseases like malaria, filariasis, yellow fever, dengue, and schistosomiasis (Bilharzia) in many areas. Health costs, appraisals of health impacts and mitigation measures are rarely part of irrigation projects, if at all.
- to mitigate the adverse effects of shallow water tables and soil salinization, some form of watertable control, soil salinity control, drainage and drainage system is needed
- as drainage water moves through the soil profile it may dissolve nutrients (either fertilizer-based or naturally occurring) such as nitrates, leading to a buildup of those nutrients in the ground-water aquifer. High nitrate levels in drinking water can be harmful to humans, particularly infants under 6 months, where it is linked to blue-baby syndrome" (see Methemoglobinemia).

Reduced Downstream River Water Quality

Owing to drainage of surface and groundwater in the project area, which waters may be salinized and polluted by agricultural chemicals like biocides and fertilizers, the quality of the river water below the project area can deteriorate, which makes it less fit for industrial, municipal and household use. It may lead to reduced public health. Polluted river water entering the sea may adversely affect the ecology along the sea shore (see Aswan dam).

The natural build up of sedimentation can reduce downstream river flows due to the installation of irrigation systems. Sedimentation is an essential part of the ecosystem that requires the natural flux of the river flow. This natural cycle of sediment dispersion replenishes the nutrients in the soil, that will in turn, determine the livelihood of the plants and animals that rely on the sediments carried downstream. The benefits of heavy deposits of sedimentation can be seen in large rivers like the Nile River. The sediment from the delta has built up to form a giant aquifer during flood season, and retains water in the wetlands. The wetlands that are created and sustained due to built up sediment at the basin of the river is a habitat for numerous species of birds However, heavy sedimentation can reduce downstream river water quality and can exacerbate floods up stream. This has been known to happen in the Sanmenxia reservoir in China. The Sanmenxia reservoir is part of a larger man-made project of hydro-electric dams called the Three Gorge Project In 1998, uncertain calculations and heavy sediment greatly affected the reservoir's ability to properly fulfill its flood-control function This also reduces the down stream river water quality. Shifting more towards mass irrigation installments in order to meet more socioeconomic demands is going against the natural balance of nature, and use water pragmatically- use it where it is found

Affected Downstream Water Users

Water becomes scarce for nomadic pastoralist in Baluchistan due to new irrigation developments.

Downstream water users often have no legal water rights and may fall victim of the development of irrigation.

Pastoralists and nomadic tribes may find their land and water resources blocked by new irrigation developments without having a legal recourse.

Flood-recession cropping may be seriously affected by the upstream interception of river water for irrigation purposes.

- In Baluchistan, Pakistan, the development of new small-scale irrigation projects depleted the water resources of nomadic tribes traveling annually between Baluchistan and Gujarat or Rajasthan, India
- After the closure of the Kainji dam, Nigeria, 50 to 70 per cent of the downstream area of flood-recession cropping was lost

Lake Manantali, 477 km², displaced 12,000 people.

Lost Land Use Opportunities

Irrigation projects may reduce the fishing opportunities of the original population and the grazing opportunities for cattle. The livestock pressure on the remaining lands may increase considerably, because the ousted traditional pastoralist tribes will have to find their subsistence and existence elsewhere, overgrazing may increase, followed by serious soil erosion and the loss of natural resources. The Manatali reservoir formed by the Manantali dam in Mali intersects the migration routes of nomadic pastoralists and destroyed 43000 ha of savannah, probably leading to overgrazing and erosion elsewhere. Further, the reservoir destroyed 120 km^2 of forest. The depletion of groundwater aquifers, which is caused by the suppression of the seasonal flood cycle, is damaging the forests downstream of the dam.

Groundwater Mining with Wells, Land Subsidence

Flooding as a consequence of land subsidence.

When more groundwater is pumped from wells than replenished, storage of water in the aquifer is being mined and the use of that water is no longer sustainable. As levels fail, it becomes more difficult to extract water and pumps will struggle to maintain the design flowrate and consume more may fenergy per unit of water. Eventually it may become so difficult to extract groundwater that farmers may be forced to abandon irrigated agriculture. Some notable examples include:

- The hundreds of tubewells installed in the state of Uttar Pradesh, India, with World Bank funding have operating periods of 1.4 to 4.7 hours/day, whereas they were designed to operate 16 hours/day
- In Baluchistan, Pakistan, the development of tubewell irrigation projects was at the expense of the traditional qanat or karez users
- Groundwater-related subsidence of the land due to mining of groundwater occurred in the United States at a rate of 1m for each 13m that the watertable was lowered
- Homes at Greens Bayou near Houston, Texas, where 5 to 7 feet of subsidence has occurred, were flooded during a storm in June 1989 as shown in the picture

Simulation and Prediction

The effects of irrigation on watertable, soil salinity and salinity of drainage and groundwater, and the effects of mitigative measures can be simulated and predicted using agro-hydro-salinity models like SaltMod and SahysMod

Case Studies

- In India 2.19 million ha have been reported to suffer from waterlogging in irrigation canal commands. Also 3.47 million ha were reported to be seriously salt affected,
- In the Indus Plains in Pakistan, more than 2 million hectares of land is waterlogged. The soil of 13.6 million hectares within the Gross Command Area was surveyed, which revealed that 3.1 million hectares (23%) was saline. 23% of this was in Sindh and 13% in the Punjab. More than 3 million ha of water-logged lands have been provided with tube-wells and drains at the cost of billions of rupees, but the reclamation objectives were only partially achieved. The Asian Development Bank (ADB) states that 38% of the irrigated area is now waterlogged and 14% of the surface is too saline for use
- In the Nile delta of Egypt, drainage is being installed in millions of hectares to combat the water-logging resulting from the introduction of massive perennial irrigation after completion of the High Dam at Assuan
- In Mexico, 15% of the 3 million ha of irrigable land is salinized and 10% is waterlogged
- In Peru some 0.3 million ha of the 1.05 million ha of irrigable land suffers from degradation (see Irrigation in Peru).
- Estimates indicate that roughly one-third of the irrigated land in the major irrigation countries is already badly affected by salinity or is expected to become so in the near future. Present estimates for Israel are 13% of the irrigated land, Australia 20%, China 15%, Iraq 50%, Egypt 30%. Irrigation-induced salinity occurs in large and small irrigation systems alike
- FAO has estimated that by 1990 about 52 million ha of irrigated land will need to have improved drainage systems installed, much of it subsurface drainage to control salinity

Reduced Downstream Drainage and Groundwater Quality

- The downstream drainage water quality may deteriorate owing to leaching of salts, nutrients, herbicides and pesticides with high salinity and alkalinity. There is threat of soils converting into saline or alkali soils. This may negatively affect the health of the population at the tail-end of the river basin and downstream of the irrigation scheme, as well as the ecological balance. The Aral Sea, for example, is seriously polluted by drainage water.

- The downstream quality of the groundwater may deteriorate in a similar way as the downstream drainage water and have similar consequences

Mitigation of Adverse Effects

Irrigation can have a variety negative impacts on ecology and socioeconomy, which may be mitigated in a number of ways. These include siting the irrigation project in a location which minimises negative impacts. The efficiency of existing projects can be improved and existing degraded croplands can be improved rather than establishing a new irrigation project Developing small-scale, individually owned irrigation systems as an alternative to large-scale, publicly owned and managed schemes. The use of sprinkler irrigation and micro-irrigation systems decrease the risk of waterlogging and erosion. Where practicable, using treated wastewater makes more water available to other users Maintaining flood flows downstream of the dams can ensure that an adequate area is flooded each year, supporting, amongst other objectives, fishery activities.

Delayed environmental impacts

It often takes time to accurately predict the impact that new irrigation schemes will have on the ecology and socioeconomy of a region. By the time these predictions are available, a considerable amount of time and resources may have already been expended in the implementation of that project. When that is the case, the project managers will often only change the project if the impact would be considerably more than they had originally expected.

Potential Benefits Outweigh the Potential Disadvantages

Frequently irrigation schemes are seen as extremely necessary for socioeconomic well-being especially in developing countries. One example of this can be demonstrated from a proposal for an irrigation scheme in Malawi. Here it was shown that the potential positive effects of the irrigation project that was being proposed "outweighed the potential negative impacts". It was stated that the impacts would mostly "be localized, minimal, short term occurring during the construction and operation phases of the Project". In order to help alleviate and prevent major environmental impacts, they would use techniques that minimize the potential negative impacts. As far as the region's socioeconomic well-being, there would be no "displacement and/or resettlement envisioned during the implementation of the Project activities". The original primary purposes of the irrigation project were to reduce poverty, improve food security, create local employment, increase household income and enhance the sustainability of land use.

Due to this careful planning this project was successful both in improving the socialeconomic conditions in the region and ensuring that land and water are sustainability into the future.

REFERENCES

1. Snyder, R. L.; Melo-Abreu, J. P. (2005). Frost protection: fundamentals, practice, and economics" (PDF). Food and Agriculture Organization of the United Nations. ISSN 1684-8241.
2. Williams, J. F.; S. R. Roberts; J. E. Hill; S. C. Scardaci; G. Tibbits. Managing Water for `Weed` Control in Rice". UC Davis, Department of Plant Sciences. Retrieved 2007-03-14.
3. Aridpoop -05-15". Retrieved 2012-06-19.
4. Hill, Donald: A History of Engineering
5. p19 Hill
6. Amenemhet III. Britannica Concise. Retrieved 2007-01-10.
7. G. Mokhtar (1981-01-01). Ancient civilizations of Africa. Unesco. International Scientific Committee for the Drafting of a General History of Africa. p. 309. ISBN 9780435948054. Retrieved 2012-06-19 – via Books.google.com.
8. Richard Bulliet, Pamela Kyle Crossley, Daniel Headrick, Steven Hirsch. Pages 53-56 (2008-06-18). The Earth and Its Peoples, Volume I: A Global History, to 1550. Books.google.com. ISBN 0618992383. Retrieved 2012-06-19.
9. Traditional technologies". Fao.org. Retrieved 2012-06-19.
10. Africa, Emerging Civilizations In Sub-Sahara Africa. Various Authors; Edited By: R. A. Guisepi". History-world.org. Retrieved 2012-06-19.
11. Dillehay TD, Eling HH Jr, Rossen J (2005). Preceramic irrigation canals in the Peruvian Andes". Proceedings of the National Academy of Sciences. **102** *(47): 17241–4. doi:10.1073/pnas.0508583102. PMC 1288011. PMID 16284247.*
12. Rodda, J. C. and Ubertini, Lucio (2004). The Basis of Civilization – Water Science? pg 161. International Association of Hydrological Sciences (International Association of Hydrological Sciences Press 2004).
13. Ancient India Indus Valley Civilization". Minnesota State University e-museum". Retrieved 2007-01-10.
14. The History of Technology – Irrigation. Encyclopædia Britannica, 1994 edition.
15. Qanat Irrigation Systems and Homegardens (Iran)". Globally Important Agriculture Heritage Systems. UN Food and Agriculture Organization. Retrieved 2007-01-10.
16. Encyclopædia Britannica, 1911 and 1989 editions
17. de Silva, Sena (1998). Reservoirs of Sri Lanka and their fisheries". UN Food and Agriculture Organization. Retrieved 2007-01-10.
18. China – history. Encyclopædia Britannica,1994 edition.
19. Needham, Joseph (1986). Science and Civilization in China: Volume 4, Physics and Physical Technology, Part 2, Mechanical Engineering. Taipei: Caves Books Ltd. Pages 344-346.
20. Needham, Volume 4, Part 2, 340-343.

21. Needham, Volume 4, Part 2, 33, 110.

22. Baek Seok-gi 백석기 (1987). Jang Yeong-sil 장영실. Woongjin Wiin Jeon-gi 웅진위인전기 11. Woongjin Publishing Co., Ltd.

23. James M. Bayman, The Hohokam of Southwest North America." Journal of World Prehistory 15.3 (2001): 257-311.

24. A new report says we`re draining our aquifers faster than ever". High Country News. 2013-06-22. Retrieved 2014-02-11.

25. Management of aquifer recharge and discharge processes and aquifer storage equilibrium" (PDF). Groundwater storage is shown to be declining in all populated continents...

26. Siebert, S.; J. Hoogeveen, P. Döll, J-M. Faurès, S. Feick, and K. Frenken (2006-11-10). The Digital Global Map of Irrigation Areas – Development and Validation of Map Version 4" (PDF). Tropentag 2006 – Conference on International Agricultural Research for Development. Bonn, Germany. Retrieved 2007-03-14.

27. The CIA World Factbook, retrieved 2011-10-30

28. Frenken, K. (2005). Irrigation in Africa in figures – AQUASTAT Survey – 2005(PDF). Food and Agriculture Organization of the United Nations. ISBN 92-5-105414-2. Retrieved 2007-03-14.

29. Provenzano, Giuseppe (2007). Using HYDRUS-2D Simulation Model to Evaluate Wetted Soil Volume in Subsurface Drip Irrigation Systems". J. Irrig. Drain Eng. **133** *(4): 342–350.* doi:10.1061/(ASCE)0733-9437(2007)133:4(342).

30. Mader, Shelli (May 25, 2010). Center pivot irrigation revolutionizesagriculture". The Fence Post Magazine. Retrieved June 6, 2012.

31. Peters, Troy. Managing Wheel - Lines and Hand - Lines for High Profitability"(PDF). Retrieved 29 May 2015.

32. Hill, Robert. Wheelmove Sprinkler Irrigation Operation and Management"(PDF). Retrieved 29 May 2015.

33. Polyester ropes natural irrigation technique". Entheogen.com. Archived from the original on April 12, 2012. Retrieved 2012-06-19.

34. Polyester rope natural irrigation technique 2". Diyrecipes.com. Retrieved 2012-06-19.

35. DIY instructions for making self-watering system using ropes". Instructables. com. 2008-03-17. Retrieved 2012-06-19.

36. Wastewater use in agriculture: Not only an issue where water is scarce!International Water Management Institute, 2010. Water Issue Brief 4

37. http://www.hydrol-earth-syst-sci.net/17/4339/2013/hess-17-4339-2013.pdf

38. http://agriwaterpedia.info/wiki/Water_use_efficiency

39. Chartres, C. and Varma, S. Out of water. From Abundance to Scarcity and How to Solve the World`s Water Problems FT Press (USA), 2010

40. Reengaging in Agricultural Water Management: Challenges and Options". The World Bank. pp. 4–5. Retrieved 2011-10-30.

41. Molden, D. (Ed). Water for food, Water for life: A Comprehensive Assessment of Water Management in Agriculture. Earthscan/IWMI, 2007.

42. ILRI, 1989, Effectiveness and Social/Environmental Impacts of Irrigation Projects: a Review. In: Annual Report 1988, International Institute for Land Reclamation and Improvement (ILRI), Wageningen, The Netherlands, pp. 18 – 34 . On line:

43. Rosegrant, Mark W., and Hans P. Binswanger. Markets in tradable water rights: potential for efficiency gains in developing country water resource allocation." World development (1994) 22#11 pp: 1613–1625.

44. EOS magazine, september 2009

45. World Water Council

46. Hukkinen, Janne, Emery Roe, and Gene I. Rochlin. A salt on the land: A narrative analysis of the controversy over irrigation-related salinity and toxicity in California`s San Joaquin Valley." Policy Sciences 23.4 (1990): 307-329. online

47. Drainage Manual: A Guide to Integrating Plant, Soil, and Water Relationships for Drainage of Irrigated Lands. Interior Dept., Bureau of Reclamation. 1993. ISBN 0-16-061623-9.

48. Free articles and software on drainage of waterlogged land and soil salinity control in irrigated land". Retrieved 2010-07-28.

49. UV treatment http://www.uvo3.co.uk/?go=Irrigation_Water

50. ultrasonic algae control http://www.lgsonic.com/irrigation-water-treatment/

CHAPTER 4

Surveying

Chapter Contents

Surveying or land surveying is the technique, profession, and science of determining the terrestrial or three-dimensional position of points and the distances and angles between them. A land surveying professional is called a land surveyor. These points are usually on the surface of the Earth, and they are often used to establish land maps and boundaries for ownership, locations like building corners or the surface location of subsurface features, or other purposes required by government or civil law, such as property sales.

INTRODUCTION

Surveyors work with elements of geometry, trigonometry, regression analysis, physics, engineering, metrology, programming languages and the law. They use equipment like total stations, robotic total stations, GPS receivers, retroreflectors, 3D scanners, radios, handheld tablets, digital levels, drones, GIS and surveying software.

A surveyor at work with an infrared reflector used for distance measurement.

Surveying has been an element in the development of the human environment since the beginning of recorded history. The planning and execution of most forms of construction require it. It is also used in transport, communications, mapping, and the definition of legal boundaries for land ownership. It is an important tool for research in many other scientific disciplines.

DEFINITIONS

ACSM

The American Congress on Surveying and Mapping (ACSM), defines surveying as *the science* and art of making all essential measurements to determine the relative position of points or physical and cultural details above, on, or beneath the surface of the Earth, and to depict them in a usable form, or to establish the position of points or details.

Also per ACSM, the type of surveying known as "land surveying" is the detailed study or inspection, as by gathering information through observations, measurements in the field, questionnaires, or research of legal instruments, and data analysis in the support of planning, designing, and establishing of property boundaries. It involves the re-establishment of cadastral surveys and land boundaries based on documents of record and historical evidence, as well as certifying surveys (as required by statute or local ordinance) of subdivision plats or maps, registered land surveys, judicial surveys, and space delineation. Land surveying can include associated services such as mapping and related data accumulation, construction layout surveys, precision measurements of length, angle, elevation, area, and volume, as well as horizontal and vertical control surveys, and the analysis and utilization of land survey data.

FIG

The International Federation of Surveyors defines the function of surveying as:

A surveyor is a professional person with the academic qualifications and technical expertise to conduct one, or more, of the following activities;

- to determine, measure and represent land, three-dimensional objects, point-fields and trajectories;
- to assemble and interpret land and geographically related information,
- to use that information for the planning and efficient administration of the land, the sea and any structures thereon; and,
- to conduct research into the above practices and to develop them.

HISTORY

Ancient Surveying

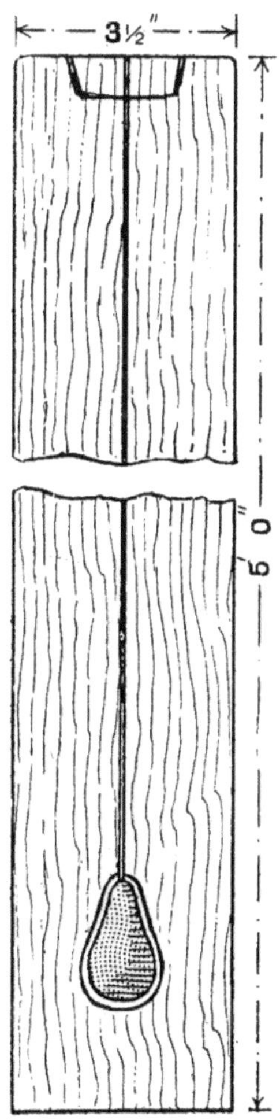

A plumb rule from the book Cassells' Carpentry and Joinery.

Basic surveyance has occurred since humans built the first large structures. The prehistoric monument at Stonehenge (c. 2500 BC) was set out by prehistoric surveyors using peg and rope geometry.

In ancient Egypt, a rope stretcher would use simple geometry to re-establish boundaries after the annual floods of the Nile River. The almost perfect squareness and north-south orientation of the Great Pyramid of Giza, built c. 2700 BC, affirm the Egyptians` command of surveying. The *Groma* instrument originated in Mesopotamia (early 1st millennium BC).

The mathematician Liu Hui described ways of measuring distant objects in his work *Haidao Suanjing* or *The Sea Island Mathematical Manual*, published in 263 AD.

The Romans recognized land surveyors as a profession. They established the basic measurements under which the Roman Empire was divided, such as a tax register of conquered lands (300 AD). Roman surveyors were known as *Gromatici*.

In medieval Europe, beating the bounds maintained the boundaries of a village or parish. This was the practice of gathering a group of residents and walking around the parish or village to establish a communal memory of the boundaries. Young boys were included to ensure the memory lasted as long as possible.

In England, William the Conqueror commissioned the Domesday Book in 1086. It recorded the names of all the land owners, the area of land they owned, the quality of the land, and specific information of the area's content and inhabitants. It did not include maps showing exact locations.

Modern Surveying

Abel Foullon described a plane table in 1551, but it is thought that the instrument was in use earlier as his description is of a developed instrument.

Gunter`s chain was introduced in 1620 by English mathematician Edmund Gunter. It enabled plots of land to be accurately surveyed and plotted for legal and commercial purposes.

Leonard Digges described a Theodolite that measured horizontal angles in his book *A geometric practice named Pantometria* (1571). Joshua Habermel (de:Erasmus Habermehl) created a theodolite with a compass and tripod in 1576. Johnathon Sission was the first to incorporate a telescope on a theodolite in 1725.

In the 18th century, modern techniques and instruments for surveying began to be used. Jesse Ramsden introduced the first precision theodolite in 1787. It was an instrument for measuring angles in the horizontal and vertical planes. He created his great theodolite using an accurate dividing engine of his own design. Ramsden's theodolite represented a great step forward in the instrument's accuracy. William Gascoigne invented an instrument that used a telescope with an installed crosshair as a target device, in 1640. James Watt developed an optical meter for the measuring of distance in 1771; it measured the parallactic angle from which the distance to a point could be deduced.

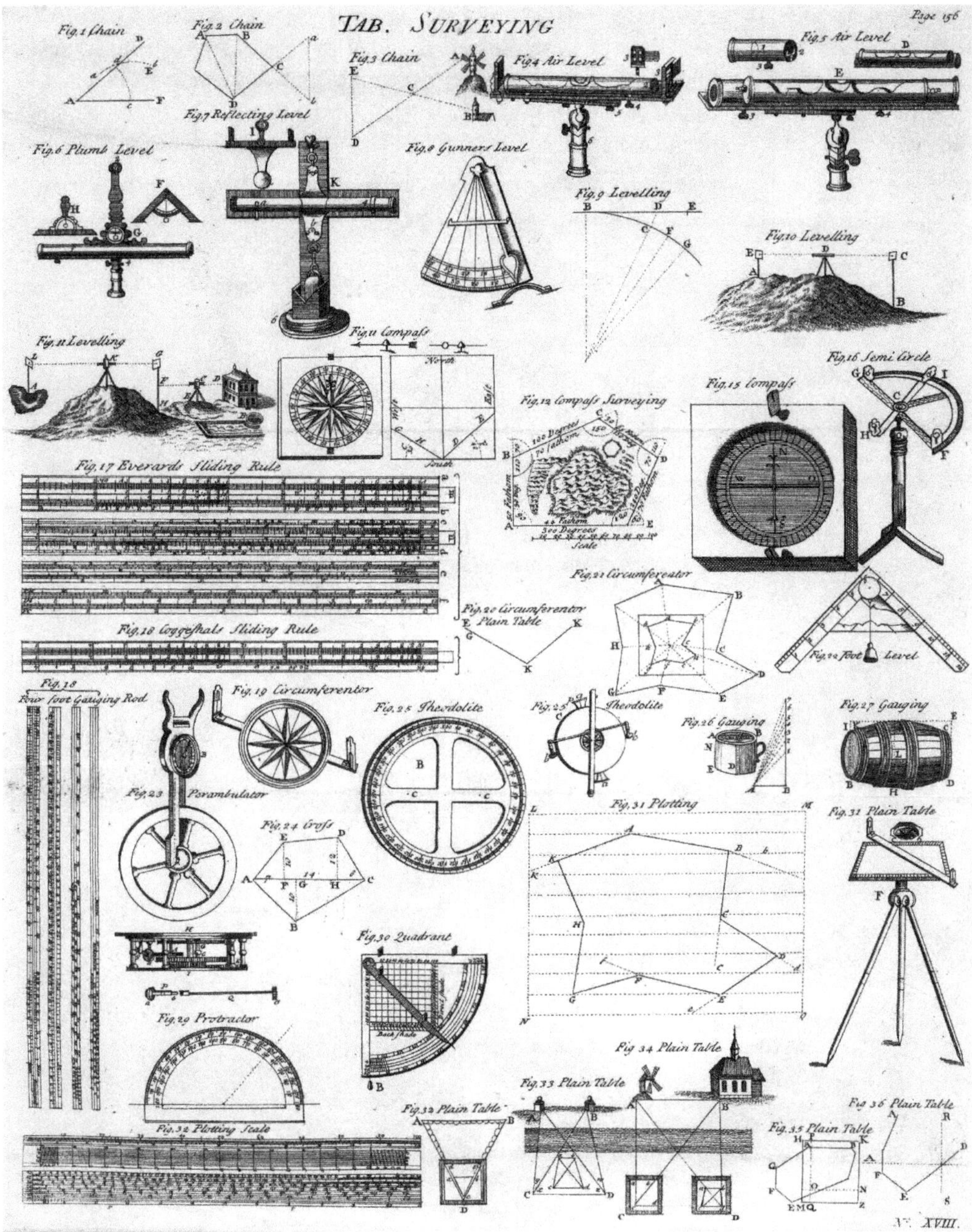

Table of Surveying, 1728 *Cyclopaedia*.

Dutch mathematician Willebrord Snellius (a.k.a. Snel van Royen) introduced the modern systematic use of triangulation. In 1615 he surveyed the distance from Alkmaar to Breda, approximately 72 miles (116,1 kilometres). He underestimated this distance by 3.5%. The survey was a chain of quadrangles containing 33 triangles in all. Snell showed how planar formulae could be corrected to allow for the curvature of the earth. He also showed how to resection, or calculate, the position of a point inside a triangle using the angles cast between the vertices at the unknown point. These could be measured more accurately than bearings of the vertices, which depended on a compass. His work established the idea of surveying a primary network of control points, and locating subsidiary points inside the primary network later. Between 1733 and 1740, Jacques Cassini and his son César undertook the first triangulation of France. They included a re-surveying of the meridian arc, leading to the publication in 1745 of the first map of France constructed on rigorous principles. By this time, triangulation methods were by then well established for local map-making.

A map of India showing the Great Trigonometrical Survey, produced in 1870.

It was only towards the end of the 18th century that detailed triangulation network surveys mapped whole countries. In 1784, a team from General William Roy's Ordnance Survey of Great Britain began the Principal Triangulation of Britain. The first Ramsden theodolite was built for this survey. The survey was finally completed in 1853. The Great Trigonometric Survey of India began in 1801. The Indian survey had an enormous scientific impact. It was responsible for one of the first accurate measurements of a section of an arc of longitude, and for measurements of the geodesic anomaly. It named and mapped Mount Everest and the other Himalayan peaks. Surveying became a professional occupation in high demand at the turn of the 19th century with the onset of the Industrial Revolution. The profession developed more accurate instruments to aid its work. Industrial infrastructure projects used surveyors to lay out canals, roads and rail.

In the US, the Land Ordinance of 1785 created the Public Land Survey System. It formed the basis for dividing the western territories into sections to allow the sale of land. The PLSS divided states into township grids which were further divided into sections and fractions of sections.

Napoleon Bonaparte founded continental Europe`s first cadastre in 1808. This gathered data on the number of parcels of land, their value, land usage, and names. This system soon spread around Europe.

A railroad surveying party at Russel`s Tank, Arizona in the 1860s

Robert Torrens introduced the Torrens system in South Australia in 1858. Torrens intended to simplify land transactions and provide reliable titles via a centralized register of land. The Torrens system was adopted in several other nations of the English-speaking world. Surveying became increasingly important with the arrival of railroads in the 1800s. Surveying was necessary so that railroads could plan technologically and financially viable routes.

20th Century

A German engineer surveying during the First World War, 1918.

At the beginning of the century surveyors had improved the older chains and ropes, but still faced the problem of accurate measurement of long distances. Dr Trevor Lloyd Wadley developed the Tellurometer during the 1950s. It measures long distances using two microwave transmitter/receivers. During the late 1950s Geodimeter introduced electronic distance measurement (EDM) equipment. EDM units use a multi frequency phase shift of light waves to find a distance. These instruments saved the need for days or weeks of chain measurement by measuring between points kilometers apart in one go.

Advances in electronics allowed miniaturization of EDM. In the 1970s the first instruments combining angle and distance measurement appeared, becoming known as total stations. Manufacturers added more equipment by degrees, bringing improvements in accuracy and speed of measurement. Major advances include tilt compensators, data recorders, and on-board calculation programs.

The first Satellite positioning system was the US Navy TRANSIT system. The first successful launch took place in 1960. The system`s main purpose was to provide position information to Polaris missile submarines. Surveyors found they could use field receivers to determine the location of a point. Sparse satellite cover and large equipment made observations laborious, and inaccurate. The main use was establishing benchmarks in remote locations.

The US Air Force launched the first prototype satellites of the Global Positioning System (GPS) in 1978. GPS used a larger constellation of satellites and improved signal transmission to provide more accuracy. Early GPS observations required several hours of observations by a static receiver to reach survey accuracy requirements. Recent improvements to both satellites and receivers allow Real Time Kinematic (RTK) surveying. RTK surveys get high-accuracy measurements by using a fixed base station and a second roving antenna. The position of the roving antenna can be tracked.

21st century

The theodolite, total station, and RTK GPS survey remain the primary methods in use.

Remote sensing and satellite imagery continue to improve and become cheaper, allowing more commonplace use. Prominent new technologies include three-dimensional (3D) scanning and use of lidar for topographical surveys. UAV technology along with photogrammetric image processing is also appearing.

SURVEYING EQUIPMENT

Hardware

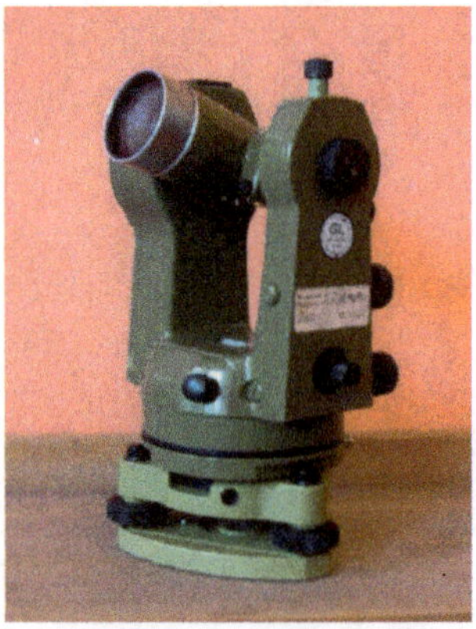

Surveying equipment. Clockwise from upper left: optical theodolite, robotic total station, RTK GPS base station, optical level.

The main surveying instruments in use around the world are the theodolite, measuring tape, total station, 3D scanners, GPS/GNSS, level and rod. Most instruments screw onto a tripod when in use. Tape measures are often used for measurement of smaller distances. 3D scanners and various forms of aerial imagery are also used.

The theodolite is an instrument for the measurement of angles. It uses two separate *circles*, *protractors* or *alidades* to measure angles in the horizontal and the vertical plane. A telescope mounted on trunnions is aligned vertically with the target object. The whole upper section rotates for horizontal alignment. The vertical circle measures the angle that the telescope makes against the vertical, known as the zenith angle. The horizontal circle uses an upper and lower plate. When beginning the survey, the surveyor points the instrument in a known direction (bearing), and clamps the lower plate in place. The instrument can then rotate to measure the bearing to other objects. If no bearing is known or direct angle measurement is wanted, the instrument can be set to zero during the initial sight. It will then read the angle between the initial object, the theodolite itself, and the item that the telescope aligns with.

The gyrotheodolite is a form of theodolite that uses a gyroscope to orient itself in the absence of reference marks. It is used in underground applications.

The total station is a development of the theodolite with an electronic distance measurement device (EDM). A total station can be used for leveling when set to the horizontal plane. Since their introduction, total stations have shifted from optical-mechanical to fully electronic devices.

Modern top-of-the-line total stations no longer need a reflector or prism to return the light pulses used for distance measurements. They are fully robotic, and can even e-mail point data to a remote computer and connect to satellite positioning systems, such as Global Positioning System. Real time kinematic GPS systems have increased the speed of surveying, but they are still only horizontally accurate to about 20 mm and vertically to 30–40 mm.

GPS surveying differs from other GPS uses in the equipment and methods used. Static GPS uses two receivers placed in position for a considerable length of time. The long span of time lets the receiver compare measurements as the satellites orbit. The changes as the satellites orbit also provide the measurement network with well conditioned geometry. This produces an accurate baseline that can be over 20 km long. RTK surveying uses one static antenna and one roving antenna. The static antenna tracks changes in the satellite positions and atmospheric conditions. The surveyor uses the roving antenna to measure the points needed for the survey. The two antennas use a radio link that allows the static antenna to send corrections to the roving antenna. The roving antenna then applies those corrections to the GPS signals it is receiving to calculate its own position. RTK surveying covers smaller distances than static methods. This is because divergent conditions further away from the base reduce accuracy.

Surveying instruments have characteristics that make them suitable for certain uses. Theodolites and levels are often used by constructors rather than surveyors in first world countries. The constructor can perform simple survey tasks using a relatively cheap instrument. Total stations are workhorses for many professional surveyors because they are versatile and reliable in all conditions. The productivity improvements from a GPS on large scale surveys makes them popular for major infrastructure or data gathering projects. One-person robotic-guided total stations allow surveyors to measure without extra workers to aim the telescope or record data. A fast but expensive way to measure large areas is with a helicopter, using a GPS to record the location of the helicopter and a laser scanner to measure the ground. To increase precision, surveyors place beacons on the ground (about 20 km (12 mi) apart). This method reaches precisions between 5–40 cm (depending on flight height).

Surveyors use ancillary equipment such as tripods and instrument stands; staves and beacons used for sighting purposes; PPE; vegetation clearing equipment; digging implements for finding survey markers buried over time; hammers for placements of markers in various surfaces and structures; and portable radios for communication over long lines of sight.

Software

Land surveyors, construction professionals and civil engineers using total station, GPS, 3D scanners and other collector data use Land Surveying Software to increase efficiency, accuracy and productivity. Land Surveying Software is a staple of contemporary land surveying.

SURVEYING TECHNIQUES

A standard Brunton Geo compass, still used commonly today by geographers, geologists and surveyors for field-based measurements.

Surveyors determine the position of objects by measuring angles and distances. The factors that can affect the accuracy of their observations are also measured. They then use this data to create vectors, bearings, co-ordinates, elevations, areas, volumes, plans and maps. Measurements are often split into horizontal and vertical components to simplify calculation. GPS and astronomic measurements also need measurement of a time component.

Distance Measurement

Example of modern equipment for surveying (Field-Map technology): GPS, laser rangefinder and field computer allows surveying as well as cartography (creation of map in real-time) and field data collection.

Before EDM devices, distances were measured using a variety of means. These included chains with links of a known length such as a Gunter`s chain, or measuring tapes made of steel or invar. To measure horizontal distances, these chains or tapes were pulled taut to reduce sagging and slack. The distance had to be adjusted for heat expansion. Attempts to hold the measuring instrument level would also be made. When measuring up a slope, the surveyor might have to break (break chain) the measurement- use an increment less than the total length of the chain. Perambulators, or measuring wheels, were used to measure longer distances but not to a high level of accuracy. Tacheometry is the science of measuring distances by measuring the angle between two ends of an object with a known size. It was sometimes used before to the invention of EDM where rough ground made chain measurement impractical.

Angle Measurement

Historically, horizontal angles were measured by using a compass to provide a magnetic bearing or azimuth. Later, more precise scribed discs improved angular resolution. Mounting telescopes with reticles atop the disc allowed more precise sighting (see theodolite). Levels and calibrated circles allowed measurement of vertical angles. Verniers allowed measurement to a fraction of a degree, such as with a turn-of-the-century transit.

The plane table provided a graphical method of recording and measuring angles, which reduced the amount of mathematics required. In 1829 Francis Ronalds invented a reflecting instrument for recording angles graphically by modifying the octant.

By observing the bearing from every vertex in a figure, a surveyor can measure around the figure. The final observation will be between the two points first observed, except with a 180° difference. This is called a *close*. If the first and last bearings are different, this shows the error in the survey, called the *angular misclose*. The surveyor can use this information to prove that the work meets the expected standards.

Levelling

Center for Operational Oceanographic Products and Services staff member conducts tide station leveling in support of the US Army Corp of Engineers in Richmond, Maine.

The simplest method for measuring height is with an altimeter using air pressure to find height. When more precise measurements are needed, means like precise levels (also known as differential leveling) are used. When precise leveling, a series of measurements between two points are taken using an instrument and a measuring rod. Differences in height between the measurements are added and subtracted in a series to get the net difference in elevation between the two endpoints. With the Global Positioning System (GPS), elevation can be measured with satellite receivers. Usually GPS is somewhat less accurate than traditional precise leveling, but may be similar over long distances.

When using an optical level, the endpoint may be out of the effective range of the instrument. There may be obstructions or large changes of elevation between the endpoints. In these situations, extra setups are needed. *Turning* is a term used when referring to moving the level to take an elevation shot from a different location. To "turn" the level, one must first take a reading and record the elevation of the point the rod is located on. While the rod is being kept in exactly the same location, the level is moved to a new location where the rod is still visible. A reading is taken from the new location of the level and the height difference is used to find the new elevation of the level gun. This is repeated until the series of measurements is completed. The level must be horizontal to get a valid measurement. Because of this, if the horizontal crosshair of the instrument is lower than the base of the rod, the surveyor will not be able to sight the rod and get a reading. The rod can usually be raised up to 25 feet high, allowing the level to be set much higher than the base of the rod.

Determining Position

The primary way of determining one's position on the earth's surface when no known positions are nearby is by astronomic observations. Observations to the sun, moon and stars could all be made using navigational techniques. Once the instrument's position and bearing to a star is determined, the bearing can be transferred to a reference point on the earth. The point can then be used as a base for further observations. Survey-accurate astronomic positions were difficult to observe and calculate and so tended to be a base off which many other measurements were made. Since the advent of the GPS system, astronomic observations are rare as GPS allows adequate positions to be determined over most of the surface of the earth.

Reference Networks

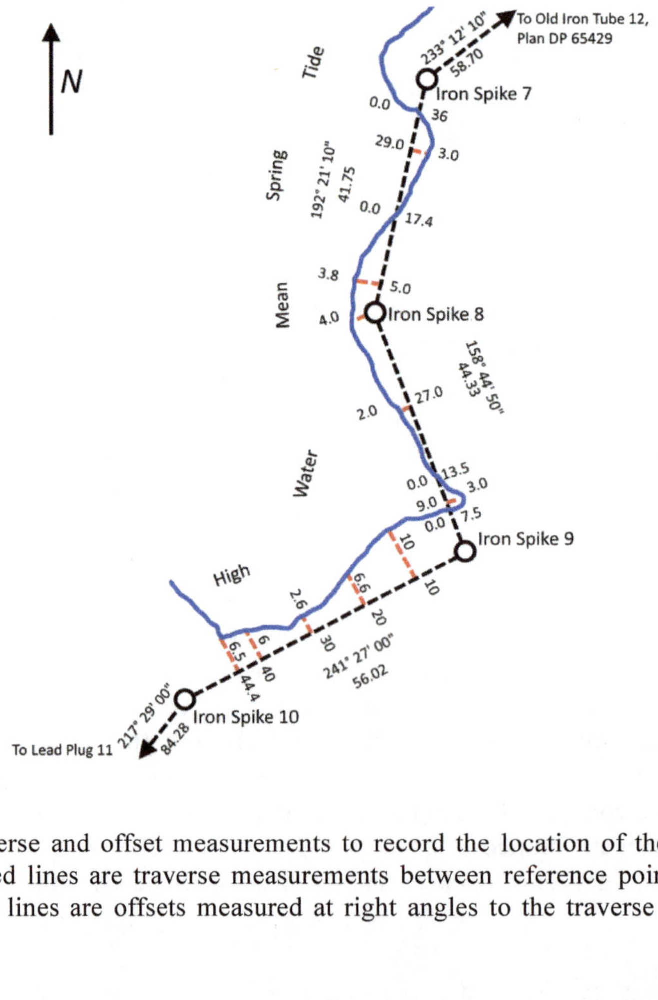

A survey using traverse and offset measurements to record the location of the shoreline shown in blue. Black dashed lines are traverse measurements between reference points (black circles). The red lines are offsets measured at right angles to the traverse lines.

Few survey positions are derived from first principles. Instead, most surveys points are measured relative to previous measured points. This forms a reference or *control* network where each point can be used by a surveyor to determine their own position when beginning a new survey.

Survey points are usually marked on the earth`s surface by objects ranging from small nails driven into the ground to large beacons that can be seen from long distances. The surveyors can set up their instruments on this position and measure to nearby objects. Sometimes a tall, distinctive feature such as a steeple or radio aerial has its position calculated as a reference point that angles can be measured against.

Triangulation is a method of horizontal location favoured in the days before EDM and GPS measurement. It can determine distances, elevations and directions between distant objects. Since the early days of surveying, this was the primary method of determining accurate positions of objects for topographic maps of large areas. A surveyor first needs to know the horizontal distance between two of the objects, known as the baseline. Then the heights, distances and angular position of other objects can be derived, as long as they are visible from one of the original objects. High-accuracy transits or theodolites were used, and angle measurements repeated for increased accuracy. See also Triangulation in three dimensions.

Offsetting is an alternate method of determining position of objects, and was often used to measure imprecise features such as riverbanks. The surveyor would mark and measure two known positions on the ground roughly parallel to the feature, and mark out a baseline between them. At regular intervals, a distance was measured at right angles from the first line to the feature. The measurements could then be plotted on a plan or map, and the points at the ends of the offset lines could be joined to show the feature.

Traversing is a common method of surveying smaller areas. The surveyor starts from an old reference mark or known position and places a network of reference marks covering the survey area. They then measure bearings and distances between the reference marks, and to the target features. Most traverses form a loop pattern or link between two prior reference marks so the surveyor can check their measurements.

Datum and Coordinate Systems

Many surveys do not calculate positions on the surface of the earth, but instead measure the relative positions of objects. However, often the surveyed items need to be compared to outside data, such as boundary lines or previous surveys objects. The oldest way of describing a position is via latitude and longitude, and often a height above sea level. As the surveying profession grew it created Cartesian coordinate systems to simplify the mathematics for surveys over small parts of the earth. The simplest coordinate systems assume that the earth is flat and measure from an arbitrary point, known as a 'datum' (singular form of data). The coordinate system allows easy calculation of the distances and direction between objects over small areas. Large areas distort due to the earth's curvature. North is often defined as true north at the datum. For larger regions, it is necessary to model the shape of the earth using an ellipsoid or a geoid. Many countries have created coordinate-grids customized to lessen error in their area of the earth.

Errors and Accuracy

A basic tenet of surveying is that no measurement is perfect, and that there will always be a small amount of error. There are three classes of survey errors:

- *Gross errors or blunders:* Errors made by the surveyor during the survey. Upsetting the instrument, misaiming a target, or writing down a wrong measurement are all gross errors. A large gross error may reduce the accuracy to an unacceptable level. Therefore, surveyors use redundant measurements and independent checks to detect these errors early in the survey.
- *Systematic:* Errors that follow a consistent pattern. Examples include effects of temperature on a chain or EDM measurement, or a poorly adjusted spirit-level causing a tilted instrument or target pole. Systematic errors that have known effects can be compensated or corrected.
- *Random:* Random errors are small unavoidable fluctuations. They are caused by imperfections in measuring equipment, eyesight, and conditions. They can be minimized by redundancy of measurement and avoiding unstable conditions. Random errors tend to cancel each other out, but checks must be made to ensure they are not propagating from one measurement to the next.

Surveyors avoid these errors by calibrating their equipment, using consistent methods, and by good design of their reference network. Repeated measurements can be averaged and any outlier measurements discarded. Independent checks like measuring a point from two or more locations or using two different methods are used. Errors can be detected by comparing the results of the two measurements.

Once the surveyor has calculated the level of the errors in his work, it is adjusted. This is the process of distributing the error between all measurements. Each observation is weighted according to how much of the total error it is likely to have caused and part of that error is allocated to it in a proportional way. The most common methods of adjustment are the Bowditch method, also known as the compass rule, and the Principle of least squares method.

The surveyor must be able to distinguish between accuracy and precision. In the United States, surveyors and civil engineers use units of feet wherein a survey foot breaks down into 10ths and 100ths. Many deed descriptions containing distances are often expressed using these units (125.25 ft). On the subject of accuracy, surveyors are often held to a standard of one one-hundredth of a foot; about 1/8 inch. Calculation and mapping tolerances are much smaller wherein achieving near-perfect closures are desired. Though tolerances will vary from project to project, in the field and day to day usage beyond a 100th of a foot is often impractical.

TYPES OF SURVEYS

Local organisations or regulatory bodies class specializations of surveying in different ways. Broad groups are:

- *As-built survey*: a survey that documents the location of recently constructed elements of a construction project. Asbuilt surveys are done for record, completion evaluation and payment purposes. An as-built survey is also known as a 'works as executed survey'. As built surveys are often presented in red or redline and laid over existing plans for comparison with design information.

- *Cadastral or boundary surveying*: a survey that establishes or re-establishes boundaries of a parcel using a legal description. It involves the setting or restoration of monuments or markers at the corners or along the lines of the parcel. These take the form of iron rods, pipes, or concrete monuments in the ground, or nails set in concrete or asphalt. The *ALTA/ACSM* Land Title Survey is a standard proposed by the American Land Title Association and the American Congress on Surveying and Mapping. It incorporates elements of the boundary survey, mortgage survey, and topographic survey.
- *Control surveying*: Control surveys establish reference points to use as starting positions for future surveys. Most other forms of surveying will contain elements of control surveying.
- *Construction surveying*
- *Deformation survey:* a survey to determine if a structure or object is changing shape or moving. First the positions of points on an object are found. A period of time is allowed to pass and the positions are then re-measured and calculated. Then a comparison between the two sets of positions is made.
- *Dimensional control survey:* This is a type of survey conducted in or on an non-level surface. Common in the oil and gas industry to replace old or damaged pipes on a like-for-like basis. The advantage of dimensional control survey is that the instrument used to conduct the survey does not need to be level. This is useful in the off-shore industry, as not all platforms are fixed and are thus subject to movement.
- *Engineering surveying*: topographic, layout, and as-built surveys associated with engineering design. They often need geodetic computations beyond normal civil engineering practice.
- *Foundation survey*: a survey done to collect the positional data on a foundation that has been poured and is cured. This is done to ensure that the foundation was constructed in the location, and at the elevation, authorized in the *plot plan*, *site plan*, or *subdivision plan*.
- *Hydrographic survey*: a survey conducted with the purpose of mapping the shoreline and bed of a body of water. Used for navigation, engineering, or resource management purposes.
- *Leveling*: either finds the elevation of a given point or establish a point at a given elevation.
- *LOMA survey*: Survey to change base flood line, removing property from a SFHA special flood hazard area.
- *Measured survey* : a building survey to produce plans of the building. such a survey may be conducted before renovation works, for commercial purpose, or at end of the construction process.

- *Mining surveying*: Mining surveying includes directing the digging of mine shafts and galleries and the calculation of volume of rock. It uses specialised techniques due to the restraints to survey geometry such as vertical shafts and narrow passages.
- *Mortgage survey:* A *mortgage survey* or *physical survey* is a simple survey that delineates land boundaries and building locations. It checks for *encroachment*, building setback restrictions and shows nearby flood zones. In many places a mortgage survey is a precondition for a mortgage loan.
- *Photographic control survey*: A survey that creates reference marks visible from the air to allow aerial photographs to be rectified.
- *Stakeout, Layout or Setout*: an element of many other surveys where the calculated or proposed position of an object is marked on the ground. This can be temporary or permanent. This is an important component of engineering and cadastral surveying.
- *Structural survey*: a detailed inspection to report upon the physical condition and structural stability of a building or structure. It highlights any work needed to maintain it in good repair.
- *Subdivision*: A boundary survey that splits a property into two or more smaller properties.
- *Topographic survey*: a survey that measures the elevation of points on a particular piece of land, and presents them as contour lines on a plot.

Plane and Geodetic Surveying

Based on the considerations and true shape of the earth, surveying is broadly classified into two types.

Plane surveying assumes the earth is flat. Curvature and spheroidal shape of the earth is neglected. In this type of surveying all triangles formed by joining survey lines are considered as plane triangles. It is employed for small survey works where errors due to the earth's shape are too small to matter. In *geodetic surveying* the curvature of the earth is taken into account while calculating reduced levels, angles, bearings and distances. This type of surveying is usually employed for large survey works. Survey works up to 100 square miles (260 square kilometers) are treated as plane and beyond that are treated as geodetic. In geodetic surveying necessary corrections are applied to reduced levels, bearings and other observations.

THE SURVEYING PROFESSION

The basic principles of surveying have changed little over the ages, but the tools used by surveyors have evolved. Engineering, especially civil engineering, often needs surveyors.

Surveyors help determine the placement of roads, railways, reservoirs, dams, pipelines, retaining walls, bridges, and buildings. They establish the boundaries of legal descriptions and political divisions. They also provide advice and data for *geographical information systems* (GIS) that record land features and boundaries.

The pundit (explorer) cartographer Nain Singh Rawat (19th century) received a Royal Geographical Society gold medal in 1876, for his efforts in exploring the Himalayas for the British.

An all-female surveying crew in Idaho, 1918.

Surveyors must have a thorough knowledge of algebra, basic calculus, geometry, and trigonometry. They must also know the laws that deal with surveys, real property, and contracts.

Most jurisdictions recognize three different levels of qualification:

Survey assistants or *chainmen* are usually unskilled workers who help the surveyor. They place target reflectors, find old reference marks, and mark points on the ground. The term 'chainman' derives from past use of measuring chains. An assistant would move the far end of the chain under the surveyor`s direction.

Survey technicians often operate survey instruments, run surveys in the field, do survey calculations, or draft plans. A technician usually has no legal authority and cannot certify his work. Not all technicians are qualified, but qualifications at the certificate or diploma level are available.

Licensed, registered, or chartered surveyors usually hold a degree or higher qualification. They are often required to pass further exams to join a professional association or to gain certifying status. Surveyors are responsible for planning and management of surveys. They have to ensure that their surveys, or surveys performed under their supervision, meet the legal standards. Many principals of surveying firms hold this status.

Informal Surveying

Not all surveys are carried out by professional surveyors. Depending on the jurisdiction and circumstances, the builders of a structure may set it out themselves. Surveyors often set out the most significant corners of a building. The builders then lay out the rest of the building themselves using simple surveying techniques.

Licensing

Licensing requirements vary with jurisdiction, and are commonly consistent within national borders. Prospective surveyors usually have to receive a degree in surveying, followed by a detailed examination of their knowledge of surveying law and principles specific to the region they wish to practice in, and undergo a period of on-the-job training or portfolio building before they are awarded a license to practise. Licensed surveyors usually receive a post nominal, which varies depending on where they qualified. The system has replaced older apprenticeship systems. A licensed land surveyor is generally required to sign and seal all plans. The state dictates the format, showing their name and registration number.

In many jurisdictions, surveyors must mark their registration number on survey monuments when setting boundary corners. Monuments take the form of capped iron rods, concrete monuments, or nails with washers.

Surveying Institutions

Most countries` governments regulate at least some forms of surveying. Their survey agencies establish regulations and standards. Standards control accuracy, surveying credentials, monumentation of boundaries and maintenance of geodetic networks. Many nations devolve this authority to regional entities or states/provinces. Cadastral surveys tend to be the most regulated because of the permanence of the work. Lot boundaries established by cadastral surveys may stand for hundreds of years without modification.

Most jurisdictions also have a form of professional institution representing local surveyors. These institutes often endorse or license potential surveyors, as well as set and enforce ethical standards. The largest institution is the International Federation of Surveyors (Abbreviated FIG, for French: *Fédération Internationale des Géomètres*). They represent the survey industry worldwide.

Surveying students with their professor at the Helsinki University of Technology in the late 19th century.

Building Surveying

Most English-speaking countries consider building surveying a distinct profession. They have their own professional associations and licensing requirements. Building surveyors focus on investigating the condition of buildings as well as legal compliance work.

Cadastral Surveying

One of the primary roles of the land surveyor is to determine the boundary of real property on the ground. The surveyor must determine where the adjoining landowners wish to put the boundary. The boundary is established in legal documents and plans prepared by attorneys, engineers, and land surveyors. The surveyor then puts monuments on the corners of the new boundary. They might also find or resurvey the corners of the property monumented by prior surveys.

Cadastral land surveyors are licensed by governments. The cadastral survey branch of the Bureau of Land Management (BLM) conducts most cadastral surveys in the United States. They consult with Forest Service, National Park Service, Army Corps of Engineers, Bureau of Indian Affairs, Fish and Wildlife Service, Bureau of Reclamation, and others. The BLM used to be known as the General Land Office (GLO).

In states organized per the Public Land Survey System (PLSS), surveyors must carry out BLM cadastral surveys under that system.

Cadastral surveyors often have to work around changes to the earth that obliterate or damage boundary monuments. When this happens, they must consider evidence that is not recorded on the title deed. This is known as extrinsic evidence.

REFERENCES

1. Snyder, R. L.; Melo-Abreu, J. P. (2005). Frost protection: fundamentals, practice, and economics (PDF). Food and Agriculture Organization of the United Nations. ISSN 1684-8241.
2. Williams, J. F.; S. R. Roberts; J. E. Hill; S. C. Scardaci; G. Tibbits. Managing Water for `Weed` Control in Rice . UC Davis, Department of Plant Sciences. Retrieved 2007-03-14.
3. Hill, Donald: A History of Engineering
4. Amenemhet III. Britannica Concise. Retrieved 2007-01-10.
5. G. Mokhtar (1981-01-01). Ancient civilizations of Africa. Unesco. International Scientific Committee for the Drafting of a General History of Africa. p. 309. ISBN 9780435948054. Retrieved 2012-06-19 – via Books.google.com.
6. Richard Bulliet, Pamela Kyle Crossley, Daniel Headrick, Steven Hirsch. Pages 53-56 (2008-06-18). The Earth and Its Peoples, Volume I: A Global History, to 1550. Books.google.com. ISBN 0618992383. Retrieved 2012-06-19.
7. Traditional technologies . Fao.org. Retrieved 2012-06-19.
8. Africa, Emerging Civilizations In Sub-Sahara Africa. Various Authors; Edited By: R. A. Guisepi . History-world.org. Retrieved 2012-06-19.
9. Dillehay TD, Eling HH Jr, Rossen J (2005). Preceramic irrigation canals in the Peruvian Andes . Proceedings of the National Academy of Sciences. 102 (47): 17241–4. doi:10.1073/pnas.0508583102. PMC 1288011. PMID 16284247.
10. Rodda, J. C. and Ubertini, Lucio (2004). The Basis of Civilization – Water Science? pg 161. International Association of Hydrological Sciences (International Association of Hydrological Sciences Press 2004).
11. Ancient India Indus Valley Civilization . Minnesota State University e-museum . Retrieved 2007-01-10.
12. The History of Technology – Irrigation. Encyclopædia Britannica, 1994 edition.
13. Qanat Irrigation Systems and Homegardens (Iran) . Globally Important Agriculture Heritage Systems. UN Food and Agriculture Organization. Retrieved 2007-01-10.
14. Encyclopædia Britannica, 1911 and 1989 editions
15. de Silva, Sena (1998). Reservoirs of Sri Lanka and their fisheries . UN Food and Agriculture Organization. Retrieved 2007-01-10.
16. China – history. Encyclopædia Britannica,1994 edition.
17. Needham, Joseph (1986). Science and Civilization in China: Volume 4, Physics and Physical Technology, Part 2, Mechanical Engineering. Taipei: Caves Books Ltd. Pages 344-346.

18. Needham, Volume 4, Part 2, 340-343.

19. Needham, Volume 4, Part 2, 33, 110.

20. Baek Seok-gi 백석기 (1987). Jang Yeong-sil 장영실. Woongjin Wiin Jeon-gi 웅진위인전기 11. Woongjin Publishing Co., Ltd.

21. James M. Bayman, The Hohokam of Southwest North America. Journal of World Prehistory 15.3 (2001): 257-311.

22. A new report says we`re draining our aquifers faster than ever . High Country News. 2013-06-22. Retrieved 2014-02-11.

23. Management of aquifer recharge and discharge processes and aquifer storage equilibrium (PDF). Groundwater storage is shown to be declining in all populated continents...

24. Siebert, S.; J. Hoogeveen, P. Döll, J-M. Faurès, S. Feick, and K. Frenken (2006-11-10). The Digital Global Map of Irrigation Areas – Development and Validation of Map Version 4 (PDF). Tropentag 2006 – Conference on International Agricultural Research for Development. Bonn, Germany. Retrieved 2007-03-14.

25. The CIA World Factbook, retrieved 2011-10-30

26. Frenken, K. (2005). Irrigation in Africa in figures – AQUASTAT Survey – 2005(PDF). Food and Agriculture Organization of the United Nations. ISBN 92-5-105414-2. Retrieved 2007-03-14.

27. Provenzano, Giuseppe (2007). Using HYDRUS-2D Simulation Model to Evaluate Wetted Soil Volume in Subsurface Drip Irrigation Systems . J. Irrig. Drain Eng. 133 (4): 342–350. doi:10.1061/(ASCE)0733-9437(2007)133:4(342).

28. Mader, Shelli (May 25, 2010). Center pivot irrigation revolutionizesagriculture . The Fence Post Magazine. Retrieved June 6, 2012.

29. Peters, Troy. Managing Wheel - Lines and Hand - Lines for High Profitability”(PDF). Retrieved 29 May 2015.

30. Hill, Robert. Wheelmove Sprinkler Irrigation Operation and Management (PDF). Retrieved 29 May 2015.

31. Polyester ropes natural irrigation technique . Entheogen.com. Archived from the original on April 12, 2012. Retrieved 2012-06-19.

32. Polyester rope natural irrigation technique 2 . Diyrecipes.com. Retrieved 2012-06-19.

33. DIY instructions for making self-watering system using ropes . Instructables. com. 2008-03-17. Retrieved 2012-06-19.

34. Wastewater use in agriculture: Not only an issue where water is scarce!International Water Management Institute, 2010. Water Issue Brief 4

35. http://www.hydrol-earth-syst-sci.net/17/4339/2013/hess-17-4339-2013.pdf

36. http://agriwaterpedia.info/wiki/Water_use_efficiency

37. Chartres, C. and Varma, S. Out of water. From Abundance to Scarcity and How to Solve the World`s Water Problems FT Press (USA), 2010

38. Reengaging in Agricultural Water Management: Challenges and Options . The World Bank. pp. 4–5. Retrieved 2011-10-30.

39. Molden, D. (Ed). Water for food, Water for life: A Comprehensive Assessment of Water Management in Agriculture. Earthscan/IWMI, 2007.

40. ILRI, 1989, Effectiveness and Social/Environmental Impacts of Irrigation Projects: a Review. In: Annual Report 1988, International Institute for Land Reclamation and Improvement (ILRI), Wageningen, The Netherlands, pp. 18 – 34 . On line: [1]

41. Rosegrant, Mark W., and Hans P. Binswanger. Markets in tradable water rights: potential for efficiency gains in developing country water resource allocation. World development (1994) 22#11 pp: 1613–1625.

42. EOS magazine, september 2009

43. World Water Council

44. Hukkinen, Janne, Emery Roe, and Gene I. Rochlin. A salt on the land: A narrative analysis of the controversy over irrigation-related salinity and toxicity in California`s San Joaquin Valley. Policy Sciences 23.4 (1990): 307-329. online

45. Drainage Manual: A Guide to Integrating Plant, Soil, and Water Relationships for Drainage of Irrigated Lands. Interior Dept., Bureau of Reclamation. 1993. ISBN 0-16-061623-9.

46. Free articles and software on drainage of waterlogged land and soil salinity control in irrigated land . Retrieved 2010-07-28.

47. UV treatment http://www.uvo3.co.uk/?go=Irrigation_Water

48. ultrasonic algae control http://www.lgsonic.com/irrigation-water-treatment/

CHAPTER 5

Seed Management Practices

Chapter Contents

A seed is a small embryonic plant enclosed in a covering called the seed coat, usually with some stored food. It is the product of the ripened ovule of gymnosperm and angiosperm plants which occurs after fertilization and some growth within the motherpl

INTRODUCTION

A **seed** is an embryonic plant enclosed in a protective outer covering. The formation of the seed is part of the process of reproduction in seed plants, the spermatophytes, including the gymnosperm and angiosperm plants.

Seeds are the product of the ripened ovule, after fertilization by pollen and some growth within the mother plant. The embryo is developed from the zygote and the seed coat from the integuments of the ovule.

Seeds have been an important development in the reproduction and success of gymnosperms and angiosperms plants, relative to more primitive plants such as ferns, mosses and liverworts, which do not have seeds and use water-dependant means to propagate themselves. Seed plants now dominate biological niches on land, from forests to grasslands both in hot and cold climates.

The term "seed" also has a general meaning that antedates the above—anything that can be sown, e.g. "seed" potatoes, "seeds" of corn or sunflower "seeds". In the case of sunflower and corn "seeds", what is sown is the seed enclosed in a shell or husk, whereas the potato is a tuber.

Many structures commonly referred to as "seeds" are actually dry fruits. Plants producing berries are called baccate. Sunflower seeds are sometimes sold commercially while still enclosed within the hard wall of the fruit, which must be split open to reach the seed. Different groups of plants have other modifications, the so-called stone fruits (such as the peach) have a hardened fruit layer (the endocarp) fused to and surrounding the actual seed. Nuts are the one-seeded, hard-shelled fruit of some plants with an indehiscent seed, such as an acorn or hazelnut.

Brown flax seeds

PRODUCTION

Seeds are produced in several related groups of plants, and their manner of production distinguishes the angiosperms ("enclosed seeds") from the gymnosperms ("naked seeds"). Angiosperm seeds are produced in a hard or fleshy structure called a fruit that encloses the seeds, hence the name. Some fruits have layers of both hard and fleshy material. In gymnosperms, no special structure develops to enclose the seeds, which begin their development "naked" on the bracts of cones. However, the seeds do become covered by the cone scales as they develop in some species of conifer.

Seed production in natural plant populations varies widely from year-to-year in response to weather variables, insects and diseases, and internal cycles within the plants themselves. Over a 20-year period, for example, forests composed of loblolly pine and shortleaf pine produced from 0 to nearly 5 million sound pine seeds per hectare. Over this period, there were six bumper, five poor, and nine good seed crops, when evaluated in regard to producing adequate seedlings for natural forest reproduction.

Development

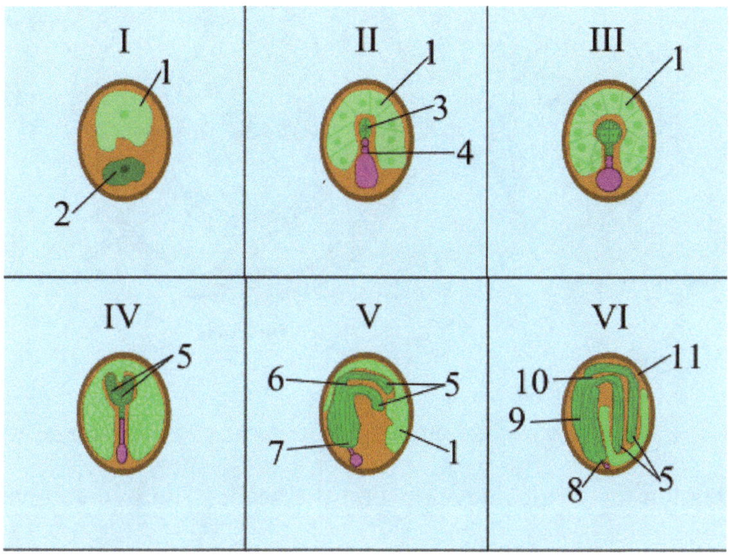

Stages of seed development:

I Zygote **II** Proembryo **III** Globular	**IV** Heart **V** Torpedo **VI** Mature Embryo

Key: *1. Endosperm 2. Zygote 3. Embryo 4. Suspensor 5. Cotyledons 6. Shoot Apical Meristem 7. Root Apical Meristem 8. Radicle 9. Hypocotyl 10. Epicotyl 11. Seed Coat*

Angiosperm (flowering plants) seeds consist of three genetically distinct constituents: (1) the embryo formed from the zygote, (2) the endosperm, which is normally triploid, (3) the seed coat from tissue derived from the maternal tissue of the ovule. In angiosperms, the process of seed development begins with double fertilization, which involves the fusion of two male gametes with the egg cell and the central cell to form the primary endosperm and the zygote. Right after fertilization, the zygote is mostly inactive, but the primary endosperm divides rapidly to form the endosperm tissue. This tissue becomes the food the young plant will consume until the roots have developed after germination.

Ovule

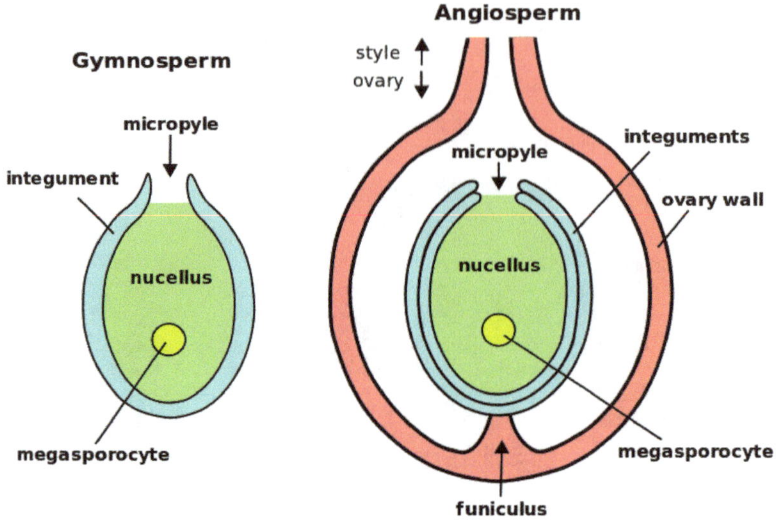

Plant ovules: Gymnosperm ovule on left, angiosperm ovule (inside ovary) on right

After fertilization the ovules develop into the seeds. The ovule consists of a number of components:

- The **funicle** (*funiculus, funiculi*) or seed stalk which attaches the ovule to the placenta and hence ovary or fruit wall, at the **pericarp**.
- The **nucellus**, the remnant of the megasporangium and main region of the ovule where the megagametophyte develops.
- The **micropyle**, a small pore or opening in the apex of the integument of the ovule where the pollen tube usually enters during the process of fertilization.
- The **chalaza**, the base of the ovule opposite the micropyle, where integument and nucellus are joined together).

The shape of the ovules as they develop often affects the final shape of the seeds. Plants generally produce ovules of four shapes: the most common shape is called anatropous, with a curved shape. Orthotropous ovules are straight with all the parts of the ovule lined up in a long row producing an uncurved seed. Campylotropous ovules have a curved megagametophyte often giving the seed a tight "C" shape. The last ovule shape is called amphitropous, where the ovule is partly inverted and turned back 90 degrees on its stalk (the funicle or funiculus).

In the majority of flowering plants, the zygote`s first division is transversely oriented in regards to the long axis, and this establishes the polarity of the embryo. The upper or chalazal pole becomes the main area of growth of the embryo, while the lower or micropylar pole produces the stalk-like suspensor that attaches to the micropyle. The suspensor absorbs and manufacturers nutrients from the endosperm that are used during the embryo`s growth.

Embryo

The inside of a *Ginkgo* seed, showing a well-developed embryo, nutritive tissue (megagametophyte), and a bit of the surrounding seed coat

The main components of the embryo are:

- The **cotyledons**, the seed leaves, attached to the embryonic axis. There may be one (Monocotyledons), or two (Dicotyledons). The cotyledons are also the source of nutrients in the non-endospermic dicotyledons, in which case they replace the endosperm, and are thick and leathery. In endospermic seeds the cotyledons are thin and papery. Dicotyledons have the point of attachment opposite one another on the axis.
- The **epicotyl**, the embryonic axis above the point of attachment of the cotyledon(s).
- The **plumule**, the tip of the epicotyl, and has a feathery appearance due to the presence of young leaf primordia at the apex, and will become the shoot upon germination.
- The **hypocotyl**, the embryonic axis below the point of attachment of the cotyledon(s), connecting the epicotyle and the radicle, being the stem-root transition zone.
- The **radicle**, the basal tip of the hypocotyl, grows into the primary root.

Monocotyledonous plants have two additional structures in the form of sheaths. The plumule is covered with a **coleoptile** that forms the first leaf while the radicle is covered with a **coleorhiza** that connects to the primary root and adventitious roots form from the sides. Here the hypocotyl is a rudimentary axis between radicle and plumule.

The seeds of corn are constructed with these structures; pericarp, scutellum (single large cotyledon) that absorbs nutrients from the endosperm, plumule, radicle, coleoptile and coleorhiza—these last two structures are sheath-like and enclose the plumule and radicle, acting as a protective covering.

Seed Coat

The maturing ovule undergoes marked changes in the integuments, generally a reduction and disorganisation but occasionally a thickening. The seed coat forms from the two integuments or outer layers of cells of the ovule, which derive from tissue from the mother plant, the inner integument forms the **tegmen** and the outer forms the **testa**. (The seed coats of some mononocotyledon plants, such as the grasses, are not distinct structures, but are fused with the fruit wall to form a pericarp.) The testae of both monocots and dicots are often marked with patterns and textured markings, or have wings or tufts of hair. When the seed coat forms from only one layer, it is also called the testa, though not all such testae are homologous from one species to the next. The funiculus abscises (detaches at fixed point – abscission zone), the scar forming an oval depression, the **hilum**. Anatropous ovules have a portion of the funiculus that is adnate (fused to the seed coat), and which forms a longitudinal ridge, or **raphe**, just above the hilum. In bitegmic ovules (e.g. *Gossypium* described here) both inner and outer integuments contribute to the seed coat formation. With continuing maturation the cells enlarge in the outer integument. While the inner epidermis may remain a single layer, it may also divide to produce two to three layers and accumulates starch, and is referred to as the colourless layer. By contrast the outer epidermis becomes tanniferous. The inner integument may consist of eight to fifteen layers. (Kozlowski 1972)

As the cells enlarge, and starch is deposited in the outer layers of the pigmented zone below the outer epidermis, this zone begins to lignify, while the cells of the outer epidermis enlarge radially and their walls thicken, with nucleus and cytoplasm compressed into the outer layer. these cells which are broader on their inner surface are called **palisade** cells. In the inner epidermis the cells also enlarge radially with plate like thickening of the walls. The mature inner integument has a palisade layer, a pigmented zone with 15-20 layers, while the innermost layer is known as the fringe layer. (Kozlowski 1972)

Gymnosperms

In gymnosperms, which do not form ovaries, the ovules and hence the seeds are exposed. This is the basis for their nomenclature – naked seeded plants. Two sperm cells transferred from the pollen do not develop the seed by double fertilization, but one sperm nucleus unites with the egg nucleus and the other sperm is not used.

Sometimes each sperm fertilizes an egg cell and one zygote is then aborted or absorbed during early development. The seed is composed of the embryo (the result of fertilization) and tissue from the mother plant, which also form a cone around the seed in coniferous plants such as pine and spruce.

Shape and Appearance

A large number of terms are used to describe seed shapes, many of which are largely self-explanatory such as *Bean-shaped* (reniform) – resembling a kidney, with lobed ends on either side of the hilum, *Square* or *Oblong* – angular with all sides more or less equal or longer than wide, *Triangular* – three sided, broadest below middle, *Elliptic* or *Ovate* or *Obovate* – rounded at both ends, or egg shaped (ovate or obovate, broader at one end), being rounded but either symmetrical about the middle or broader below the middle or broader above the middle.

Other less obvious terms include discoid (resembling a disc or plate, having both thickness and parallel faces and with a rounded margin), ellipsoid, globose (spherical), or subglobose (Inflated, but less than spherical), lenticular, oblong, ovoid, reniform and sectoroid. Striate seeds are striped with parallel, longitudinal lines or ridges. The commonest colours are brown and black, other colours are infrequent. The surface varies from highly polished to considerably roughened. The surface may have a variety of appendages (see Seed coat). A seed coat with the consistency of cork is referred to as suberose. Other terms include crustaceous (hard, thin or brittle).

Structure

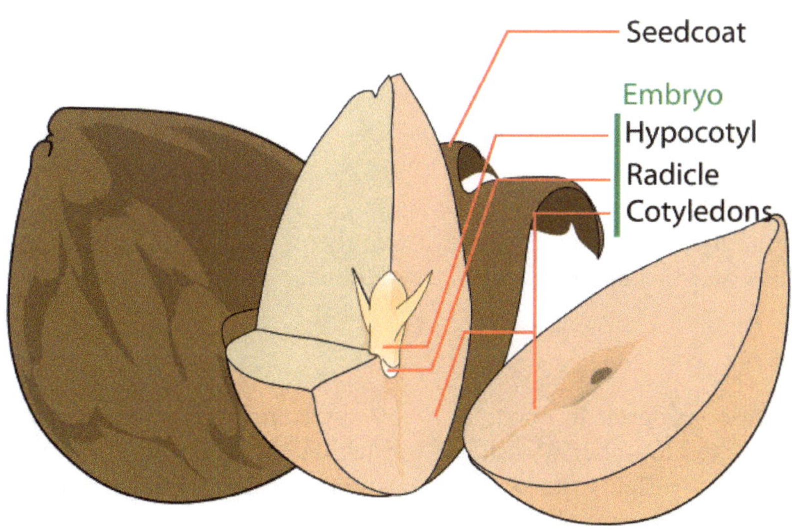

The parts of an avocado seed (a dicot), showing the seed coat and embryo

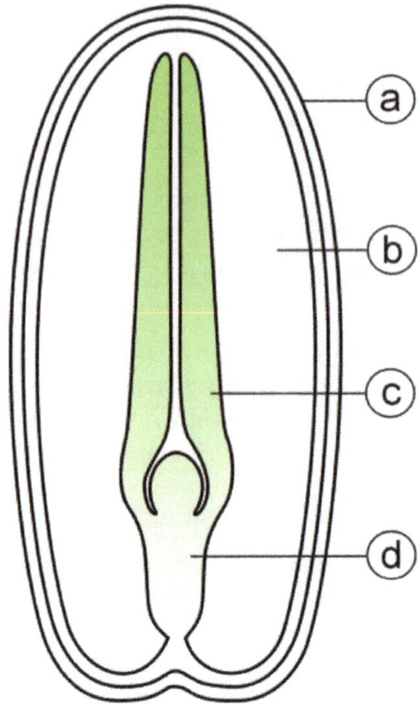

Diagram of the internal structure of a dicot seed and embryo: (a) seed coat, (b) endosperm, (c) cotyledon, (d) hypocotyl

A typical seed includes two basic parts:

- an embryo;
- a seed coat.

In addition, the **endosperm** forms a supply of nutrients for the embryo in most monocotyledons and the endospermic dicotyledons.

Seed Types

Seeds have been considered to occur in many structurally different types (Martin 1946). These are based on a number of criteria, of which the dominant one is the embryo-to-seed size ratio. This reflects the degree to which the developing cotyledons absorb the nutrients of the endosperm, and thus obliterate it.

Six types occur amongst the monocotyledons, ten in the dicotyledons, and two in the gymnosperms (linear and spatulate). This classification is based on three characteristics: embryo morphology, amount of endosperm and the position of the embryo relative to the endosperm.

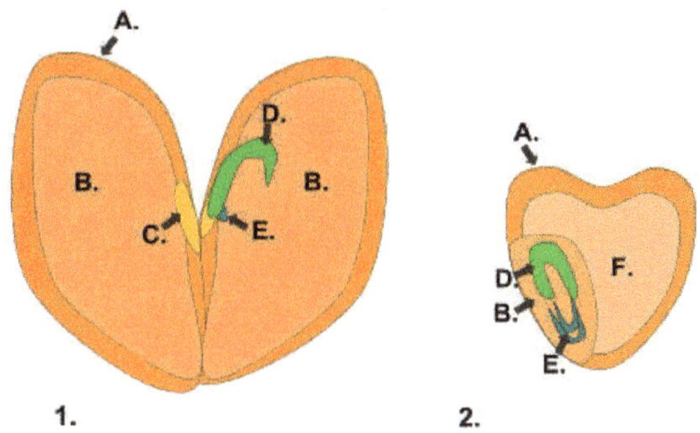

Diagram of a generalized dicot seed (1) versus a generalized monocot seed (2). A. Seed Coat B. Cotyledon C. Hilum D. Plumule E. Radicle F. Endosperm

Embryo

In endospermic seeds, there are two distinct regions inside the seed coat, an upper and larger endosperm and a lower smaller embryo. The embryo is the fertilised ovule, an immature plant from which a new plant will grow under proper conditions. The embryo has one cotyledon or seed leaf in monocotyledons, two cotyledons in almost all dicotyledons and two or more in gymnosperms. In the fruit of grains (caryopses) the single monocotyledon is shield shaped and hence called a **scutellum**. The scutellum is pressed closely against the endosperm from which it absorbs food, and passes it to the growing parts. Embryo descriptors include small, straight, bent, curved and curled.

Nutrient Storage

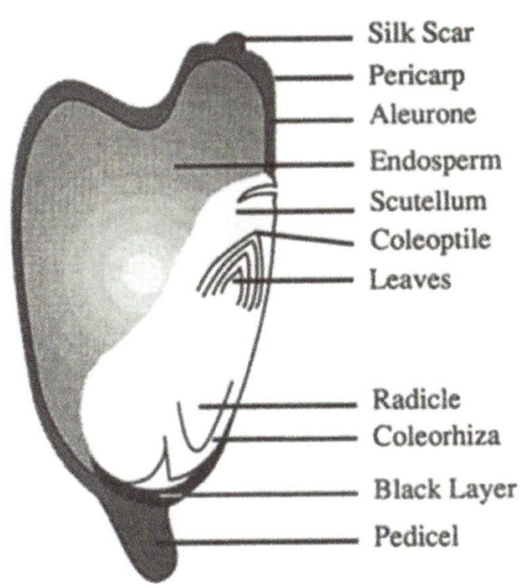

Layers within an endospermic maize seed

Within the seed, there usually is a store of nutrients for the seedling that will grow from the embryo. The form of the stored nutrition varies depending on the kind of plant. In angiosperms, the stored food begins as a tissue called the endosperm, which is derived from the mother plant and the pollen via double fertilization. It is usually triploid, and is rich in oil or starch, and protein. In gymnosperms, such as conifers, the food storage tissue (also called endosperm) is part of the female gametophyte, a haploid tissue. The endosperm is surrounded by the aleurone layer (peripheral endosperm), filled with proteinaceous aleurone grains.

Originally, by analogy with the animal ovum, the outer nucellus layer (perisperm) was referred to as albumen, and the inner endosperm layer as vitellus. Although misleading, the term began to be applied to all the nutrient matter. This terminology persists in referring to endospermic seeds as `albuminous`. The nature of this material is used in both describing and classifying seeds, in addition to the embryo to endosperm size ratio. The endosperm may be considered to be farinaceous (or mealy) in which the cells are filled with starch, as for instance cereal grains, or not (non-farinaceous). The endosperm may also be referred to as `fleshy` or `cartilaginous` with thicker soft cells such as coconut, but may also be oily as in *Ricinus* (castor oil), *Croton* and Poppy. The endosperm is called 'horny' when the cell walls are thicker such as date and coffee, or 'ruminated' if mottled, as in nutmeg, palms and Annonaceae.

In most monocotyledons (such as grasses and palms) and some (**endospermic** or **albuminous**) dicotyledons (such as castor beans) the embryo is embedded in the endosperm (and nucellus, which the seedling will use upon germination. In the **non-endospermic** dicotyledons the endosperm is absorbed by the embryo as the latter grows within the developing seed, and the cotyledons of the embryo become filled with stored food. At maturity, seeds of these species have no endosperm and are also referred to as **exalbuminous** seeds. The exalbuminous seeds include the legumes (such as beans and peas), trees such as the oak and walnut, vegetables such as squash and radish, and sunflowers. According to Bewley and Black (1978), Brazil nut storage is in hypocotyl, this place of storage is uncommon among seeds. All gymnosperm seeds are albuminous.

Seed Coat

The seed coat develops from the maternal tissue, the integuments, originally surrounding the ovule. The seed coat in the mature seed can be a paper-thin layer (e.g. peanut) or something more substantial (e.g. thick and hard in honey locust and coconut), or fleshy as in the sarcotesta of pomegranate. The seed coat helps protect the embryo from mechanical injury, predators and drying out. Depending on its development, the seed coat is either **bitegmic** or **unitegmic**. Bitegmic seeds form a testa from the outer integument and a tegmen from the inner integument while unitegmic seeds have only one integument. Usually parts of the testa or tegmen form a hard protective mechanical layer. The mechanical layer may prevent water penetration and germination. Amongst the barriers may be the presence of lignified sclereids.

The outer integument has a number of layers, generally between four and eight organised into three layers: (a) outer epidermis, (b) outer pigmented zone of two to five layers containing tannin and starch, and (c) inner epidermis. The **endotegmen** is derived from the inner epidermis of the inner integument, the **exotegmen** from the outer surface of the inner integument. The **endotesta** is derived from the inner epidermis of the outer integument, and the outer layer of the testa from the outer surface of the outer integument is referred to as the **exotesta**. If the exotesta is also the mechanical layer, this is called an exotestal seed, but if the mechanical layer is the endotegmen, then the seed is endotestal. The exotesta may consist of one or more rows of cells that are elongated and pallisade like (e.g. Fabaceae), hence 'palisade exotesta'.

In addition to the three basic seed parts, some seeds have an appendage, an **aril**, a fleshy outgrowth of the funicle (funiculus), (as in yew and nutmeg) or an oily appendage, an **elaiosome** (as in *Corydalis*), or hairs (trichomes). In the latter example these hairs are the source of the textile crop cotton. Other seed appendages include the raphe (a ridge), wings, **caruncles** (a soft spongy outgrowth from the outer integument in the vicinity of the micropyle), spines, or tubercles.

A scar also may remain on the seed coat, called the **hilum**, where the seed was attached to the ovary wall by the funicle. Just below it is a small pore, representing the micropyle of the ovule.

Size and Seed Set

A collection of various vegetable and herb seeds

Seeds are very diverse in size. The dust-like orchid seeds are the smallest, with about one million seeds per gram; they are often embryonic seeds with immature embryos and no significant energy reserves. Orchids and a few other groups of plants are mycoheterotrophs which depend on mycorrhizal fungi for nutrition during germination and the early growth of the seedling. Some terrestrial orchid seedlings, in fact, spend the first few years of their lives deriving energy from the fungi and do not produce green leaves. At over 20 kg, the largest seed is the *coco de mer*. Plants that produce smaller seeds can generate many more seeds per flower, while plants with larger seeds invest more resources into those seeds and normally produce fewer seeds. Small seeds are quicker to ripen and can be dispersed sooner, so fall blooming plants often have small seeds. Many annual plants produce great quantities of smaller seeds; this helps to ensure at least a few will end in a favorable place for growth. Herbaceous perennials and woody plants often have larger seeds; they can produce seeds over many years, and larger seeds have more energy reserves for germination and seedling growth and produce larger, more established seedlings after germination.

Functions

Seeds serve several functions for the plants that produce them. Key among these functions are nourishment of the embryo, dispersal to a new location, and dormancy during unfavorable conditions. Seeds fundamentally are means of reproduction, and most seeds are the product of sexual reproduction which produces a remixing of genetic material and phenotype variability on which natural selection acts.

Embryo Nourishment

Seeds protect and nourish the embryo or young plant. They usually give a seedling a faster start than a sporeling from a spore, because of the larger food reserves in the seed and the multicellularity of the enclosed embryo.

SEED DISPERSAL

Seed dispersal is the movement or transport of seeds away from the parent plant. Plants have very limited mobility and consequently rely upon a variety of dispersal vectors to transport their propagules, including both abiotic and biotic vectors. Seeds can be dispersed away from the parent plant individually or collectively, as well as dispersed in both space and time. The patterns of seed dispersal are determined in large part by the dispersal mechanism and this has important implications for the demographic and genetic structure of plant populations, as well as migration patterns and species interactions. There are five main modes of seed dispersal: gravity, wind, ballistic, water, and by animals. Some plants are serotinous and only disperse their seeds in response to an environmental stimulus.

Benefits

Seed dispersal is likely to have several benefits for plant species. First, seed survival is often higher away from the parent plant. This higher survival may result from the actions of density-dependent seed and seedling predators and pathogens, which often target the high concentrations of seeds beneath adults. Competition with adult plants may also be lower when seeds are transported away from their parent.

Seed dispersal also allows plants to reach specific habitats that are favorable for survival, a hypothesis known as directed dispersal. For example, *Ocotea endresiana*

(Lauraceae) is a tree species from Latin America which is dispersed by several species of birds, including the three-wattled bellbird. Male bellbirds perch on dead trees in order to attract mates, and often defecate seeds beneath these perches where the seeds have a high chance of survival because of high light conditions and escape from fungal pathogens. In the case of fleshy-fruited plants, seed-dispersal in animal guts (endozoochory) often enhances the amount, the speed, and the asynchrony of germination, which can have important plant benefits.

Seeds dispersed by ants (myrmecochory) are not only dispersed short distances but are also buried underground by the ants. These seeds can thus avoid adverse environmental effects such as fire or drought, reach nutrient-rich microsites and survive longer than other seeds. These features are peculiar to myrmecochory, which may thus provide additional benefits not present in other dispersal modes.

Finally, at another scale, seed dispersal may allow plants to colonize vacant habitats and even new geographic regions.

Types

Seed dispersal is sometimes split into *autochory* (when dispersal is attained using the plant's own means) and *allochory* (when obtained through external means).

Autochory

Autochorous plants disperse their seed without any help from an external vector, as a result this limits plants considerably as to the distance they can disperse their seed. Two other types of autochory not described in detail here are **blastochory**, where the stem of the plant crawls along the ground to deposit its seed far from the base of the plant, and **herpochory** (the seed crawls by means of trichomes and changes in humidity).

Gravity

Barochory or the plant use of gravity for dispersal is a simple means of achieving seed dispersal. The effect of gravity on heavier fruits causes them to fall from the plant when ripe. Fruits exhibiting this type of dispersal include apples, coconuts and passionfruit and those with harder shells (which often roll away from the plant to gain more distance). Gravity dispersal also allows for later transmission by water or animal.

Ballistic Dispersal

Ballochory is a type of dispersal where the seed is forcefully ejected by explosive dehiscence of the fruit. Often the force that generates the explosion results from turgor pressure within the fruit or due to internal tensions within the fruit. Some examples of plants which disperse their seeds autochorously include: *Impatiens spp.*, *Arceuthobium spp.*, *Ecballium spp.*, *Geranium spp.*, *Cardamine hirsuta* and others. An exceptional example of ballochory is *Hura crepitans*—this plant is commonly called the dynamite tree due to the sound of the fruit exploding. The explosions are powerful enough to throw the seed up to 100 meters.

Allochory

Allochory refers to any of many types of seed dispersal where a vector or secondary agent is used to disperse seeds. This vectors may include wind, water, animals or others.

Wind

Wind dispersal of dandelion seeds

Entada phaseoloides – Hydrochory

Wind dispersal (*anemochory*) is one of the more primitive means of dispersal. Wind dispersal can take on one of two primary forms: seeds can float on the breeze or alternatively, they can flutter to the ground. The classic examples of these dispersal mechanisms, in the temperate northern hemisphere, include dandelions, which have a feathery pappus attached to their seeds and can be dispersed long distances, and maples, which have winged seeds (samaras) and flutter to the ground. An important constraint on wind dispersal is the need for abundant seed production to maximize the likelihood of a seed landing in a site suitable for germination. There are also strong evolutionary constraints on this dispersal mechanism. For instance, Cody and Overton (1996) found that species in the Asteraceae on islands tended to have reduced dispersal capabilities (i.e., larger seed mass and smaller pappus) relative to the same species on the mainland. Also, *Helonias bullata*, a species perennial herb native to the United States, evolved to utilize wind dispersal as the primary seed dispersal mechanism; however, limited wind in its habitat prevents the seeds to successfully disperse away from its parents, resulting in clusters of population. Reliance on wind dispersal is common among many weedy or ruderal species. Unusual mechanisms of wind dispersal include tumbleweeds, where the entire plant is blown by the wind. Physalis fruits, when not fully ripe, may sometimes be dispersed by wind due to the space between the fruit and the covering calyx which acts as air bladder.

Water

Many aquatic (water dwelling) and some terrestrial (land dwelling) species use *hydrochory*, or seed dispersal through water. Seeds can travel for extremely long distances, depending on the specific mode of water dispersal; this especially applies to fruits which are waterproof and float.

The water lily is an example of such a plant. Water lilies' flowers make a fruit that floats in the water for a while and then drops down to the bottom to take root on the floor of the pond. The seeds of palm trees can also be dispersed by water. If they grow near oceans, the seeds can be transported by ocean currents over long distances, allowing the seeds to be dispersed as far as other continents.

Mangrove trees grow directly out of the water; when their seeds are ripe they fall from the tree and grow roots as soon as they touch any kind of soil. During low tide, they might fall in soil instead of water and start growing right where they fell. If the water level is high, however, they can be carried far away from where they fell. Mangrove trees often make little islands as dirt and other things collect in their roots, making little bodies of land.

A special review for oceanic waters hydrochory can be seen at oceanic dispersal.

Animals can disperse plant seeds in several ways, all named *zoochory*. Seeds can be transported on the outside of vertebrate animals (mostly mammals), a process known as *epizoochory*. Plant species transported externally by animals can have a variety of adaptations for dispersal, including adhesive mucus, and a variety of hooks, spines and barbs. A typical example of an epizoochorous plant is *Trifolium angustifolium*, a species of Old World clover which adheres to animal fur by means of stiff hairs covering the seed. Epizoochorous plants tend to be herbaceous plants, with many representative species in the families Apiaceae and Asteraceae.However, epizoochory is a relatively rare dispersal syndrome for plants as a whole; the percentage of plant species with seeds adapted for transport on the outside of animals is estimated to be below 5%. Nevertheless, epizoochorous transport can be highly effective if seeds attach to wide-ranging animals. This form of seed dispersal has been implicated in rapid plant migration and the spread of invasive species.

The "bill" and seed dispersal mechanism of *Geranium pretense*

By Animals

The small hooks on the surface of a bur enable attachment to animal fur for dispersion.

Seed dispersal via ingestion by vertebrate animals (mostly birds and mammals), or *endozoochory*, is the dispersal mechanism for most tree species. Endozoochory is generally a coevolved mutualistic relationship in which a plant surrounds seeds with an edible, nutritious fruit as a good food for animals that consume it. Birds and mammals are the most important seed dispersers, but a wide variety of other animals, including turtles and fish, can transport viable seeds. The exact percentage of tree species dispersed by endozoochory varies between habitats, but can range to over 90% in some tropical rainforests. Seed dispersal by animals in tropical rainforests has received much attention, and this interaction is considered an important force shaping the ecology and evolution of vertebrate and tree populations. In the tropics, large animal seed dispersers (such as tapirs, chimpanzees and hornbills) may disperse large seeds with few other seed dispersal agents. The extinction of these large frugivores from poaching and habitat loss may have negative effects on the tree populations that depend on them for seed dispersal.

Seed dispersal by ants (*myrmecochory*) is a dispersal mechanism of many shrubs of the southern hemisphere or understorey herbs of the northern hemisphere.Seeds of myrmecochorous plants have a lipid-rich attachment called the elaiosome, which attracts ants. Ants carry such seeds into their colonies, feed the elaiosome to their larvae and discard the otherwise intact seed in an underground chamber. Myrmecochory is thus a coevolved mutualistic relationship between plants and seed-disperser ants. Myrmecochory has independently evolved at least 100 times in flowering plants and is estimated to be present in at least 11 000 species, but likely up to 23 000 or 9% of all species of flowering plants. Myrmecochorous plants are most frequent in the fynbos vegetation of the Cape Floristic Region of South Africa, the kwongan vegetation and other dry habitat types of Australia, dry forests and grasslands of the Mediterranean region and northern temperate forests of western Eurasia and eastern North America, where up to 30–40% of understorey herbs are myrmecochorous.

Seed predators, which include many rodents (such as squirrels) and some birds (such as jays) may also disperse seeds by hoarding the seeds in hidden caches.The seeds in caches are usually well-protected from other seed predators and if left uneaten will grow into new plants. In addition, rodents may also disperse seeds via seed spitting due to the presence of secondary metabolites in ripe fruits. Finally, seeds may be secondarily dispersed from seeds deposited by primary animal dispersers. For example, dung beetles are known to disperse seeds from clumps of feces in the process of collecting dung to feed their larvae.

Other types of zoochory are *chiropterochory* (by bats), *malacochory* (by molluscs, mainly terrestrial snails), *ornithochory* (by birds) and *saurochory* (by non-bird sauropsids).

By Humans

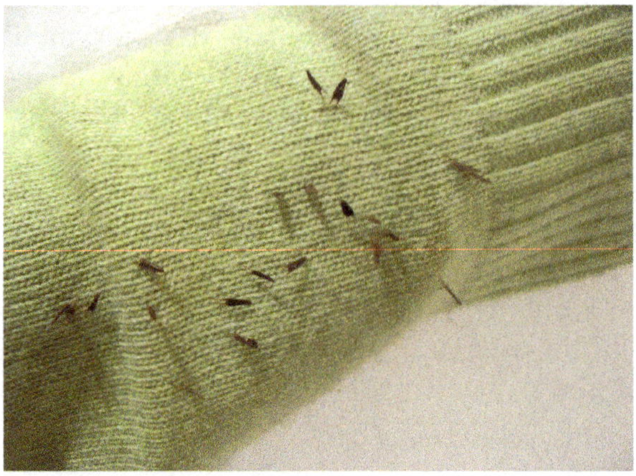

Epizoochory in *Bidens tripartita*; the seeds have attached to the clothes of a human.

Seed dispersal by a car

Dispersal by humans (*anthropochory*) used to be seen as a form of dispersal by animals. Recent research points out that human dispersers differ from animal dispersers by a much higher mobility based on the technical means of human transport. Dispersal by humans on the one hand may act on large geographical scales and lead to invasive species. On the other hand, dispersal by humans also acts on smaller, regional scales and drives the dynamics of existing biological populations. Humans may disperse seeds by many various means and some surprisingly high distances have been repeatedly measured. Examples are: dispersal on human clothes (up to 250 m), on shoes (up to 5 km) or by cars (regularly ~ 250 m, singles cases > 100 km).

Deliberate seed dispersal also occurs as seed bombing. This has risks as unsuitable provenance may introduce genetically unsuitable plants to new environments.

Consequences

Seed dispersal has many consequences for the ecology and evolution of plants. Dispersal is necessary for species migrations, and in recent times dispersal ability is an important factor in whether or not a species transported to a new habitat by humans will become an invasive species. Dispersal is also predicted to play a major role in the origin and maintenance of species diversity. For example, myrmecochory increased the rate of diversification more than twofold in plant groups in which it has evolved because myrmecochorous lineages contain more than twice as many species as their non-myrmecochorous sister groups. Dispersal of seeds away from the parent organism has a central role in two major theories for how biodiversity is maintained in natural ecosystems, the Janzen-Connell hypothesis and recruitment limitation. Seed dispersal is essential in allowing forest migration of flowering plants.

In addition, the speed and direction of wind are highly influential in the dispersal process and in turn the deposition patterns of floating seeds in the stagnant water bodies. The transportation of seeds is led by the wind direction. This effects colonization situated on the banks of a river or to wetlands adjacent to streams relative to the distinct wind directions. The wind dispersal process can also effect connections between water bodies. Essentially, wind plays a larger role in the dispersal of waterborne seeds in a short period of time, days and seasons, but the ecological process allows the process to become balanced throughout a time period of several years. The time period of which the dispersal occurs is essential when considering the consequences of wind on the ecological process.

By Wind (Anemochory)

Dandelion seeds are contained within achenes, which can be carried long distances by the wind.

The seed pod of milkweed (*Asclepias syriaca*)

- Some seeds (e.g., pine) have a wing that aids in wind dispersal.
- The dustlike seeds of orchids are carried efficiently by the wind.
- Some seeds (e.g. milkweed, poplar) have hairs that aid in wind dispersal.

Other seeds are enclosed in fruit structures that aid wind dispersal in similar ways:

- Dandelion achenes have hairs.
- Maple samaras have two wings.

By Water (Hydrochory)

- Some plants, such as *Mucuna* and *Dioclea*, produce buoyant seeds termed sea-beans or drift seeds because they float in rivers to the oceans and wash up on beaches.

By Animals (Zoochory)

- Seeds (burrs) with barbs or hooks (e.g. acaena, burdock, dock) which attach to animal fur or feathers, and then drop off later.
- Seeds with a fleshy covering (e.g. apple, cherry, juniper) are eaten by animals (birds, mammals, reptiles, fish) which then disperse these seeds in their droppings.
- Seeds (nuts) are attractive long-term storable food resources for animals (e.g. acorns, hazelnut, walnut); the seeds are stored some distance from the parent plant, and some escape being eaten if the animal forgets them.

Myrmecochory is the dispersal of seeds by ants. Foraging ants disperse seeds which have appendages called elaiosomes (e.g. bloodroot, trilliums, acacias, and many species of Proteaceae). Elaiosomes are soft, fleshy structures that contain nutrients for animals that eat them. The ants carry such seeds back to their nest, where the elaiosomes are eaten. The remainder of the seed, which is hard and inedible to the ants, then germinates either within the nest or at a removal site where the seed has been discarded by the ants. This dispersal relationship is an example of mutualism, since the plants depend upon the ants to disperse seeds, while the ants depend upon the plants seeds for food. As a result, a drop in numbers of one partner can reduce success of the other. In South Africa, the Argentine ant (*Linepithema humile*) has invaded and displaced native species of ants. Unlike the native ant species, Argentine ants do not collect the seeds of *Mimetes cucullatus* or eat the elaiosomes. In areas where these ants have invaded, the numbers of *Mimetes* seedlings have dropped.

Dormancy

Seed dormancy has two main functions: the first is synchronizing germination with the optimal conditions for survival of the resulting seedling; the second is spreading germination of a batch of seeds over time so a catastrophe (e.g. late frosts, drought, herbivory) does not result in the death of all offspring of a plant (bet-hedging). Seed dormancy is defined as a seed failing to germinate under environmental conditions optimal for germination, normally when the environment is at a suitable temperature with proper soil moisture. This true dormancy or innate dormancy is therefore caused by conditions within the seed that prevent germination. Thus dormancy is a state of the seed, not of the environment. Induced dormancy, enforced dormancy or seed quiescence occurs when a seed fails to germinate because the external environmental conditions are inappropriate for germination, mostly in response to conditions being too dark or light, too cold or hot, or too dry.

Seed dormancy is not the same as seed persistence in the soil or on the plant, though even in scientific publications dormancy and persistence are often confused or used as synonyms.

Often, seed dormancy is divided into four major categories: exogenous; endogenous; combinational; and secondary. A more recent system distinguishes five classes: morphological, physiological, morphophysiological, physical and combinational dormancy.

Exogenous dormancy is caused by conditions outside the embryo, including:

- **Physical dormancy** or **hard seed coats** occurs when seeds are impermeable to water. At dormancy break, a specialized structure, the 'water gap', is disrupted in response to environmental cues, especially temperature, so water can enter the seed and germination can occur. Plant families where physical dormancy occurs include Anacardiaceae, Cannaceae, Convulvulaceae, Fabaceae and Malvaceae.
- **Chemical dormancy** considers species that lack physiological dormancy, but where a chemical prevents germination. This chemical can be leached out of the seed by rainwater or snow melt or be deactivated somehow. Leaching of chemical inhibitors from the seed by rain water is often cited as an important cause of dormancy release in seeds of desert plants, but little evidence exists to support this claim.

Endogenous dormancy is caused by conditions within the embryo itself, including:

- In **morphological dormancy**, germination is prevented due to morphological characteristics of the embryo. In some species, the embryo is just a mass of cells when seeds are dispersed; it is not differentiated. Before germination can take place, both differentiation and growth of the embryo have to occur. In other species, the embryo is differentiated but not fully grown (underdeveloped) at dispersal, and embryo growth up to a species specific length is required before germination can occur. Examples of plant families where morphological dormancy occurs are Apiaceae, Cycadaceae, Liliaceae, Magnoliaceae and Ranunculaceae.
- **Morphophysiological dormancy** includes seeds with underdeveloped embryos, and also have physiological components to dormancy. These seeds, therefore, require a dormancy-breaking treatments, as well as a period of time to develop fully grown embryos. Plant families where morphophysiological dormancy occurs include Apiaceae, Aquifoliaceae, Liliaceae, Magnoliaceae, Papaveraceae and Ranunculaceae. Some plants with morphophysiological dormancy, such as *Asarum* or *Trillium* species, have multiple types of dormancy, one affects radicle (root) growth, while the other affects plumule (shoot) growth. The terms "double dormancy" and "two-year seeds" are used for species whose seeds need two years to complete germination or at least two winters and one summer. Dormancy of the radicle (seedling root)is broken during the first winter after dispersal while dormancy of the shoot bud is broken during the second winter.
- **Physiological dormancy** means the embryo, due to physiological causes, cannot generate enough power to break through the seed coat, endosperm or other covering structures. Dormancy is typically broken at cool wet, warm wet or warm dry conditions. Abscisic acid is usually the growth inhibitor in seeds, and its production can be affected by light.
- **Drying**, in some plants, including a number of grasses and those from seasonally arid regions, is needed before they will germinate. The seeds are released, but need to have a lower moisture content before germination can begin. If the seeds remain moist after dispersal, germination can be delayed for many months or even years. Many herbaceous plants from temperate climate zones have physiological dormancy that disappears with drying of the seeds. Other species will germinate after dispersal only under very narrow temperature ranges, but as the seeds dry, they are able to germinate over a wider temperature range.
- In seeds with **combinational dormancy**, the seed or fruit coat is impermeable to water and the embryo has physiological dormancy. Depending on the species, physical dormancy can be broken before or after physiological dormancy is broken.
- **Secondary dormancy*** is caused by conditions after the seed has been dispersed and occurs in some seeds when nondormant seed is exposed to conditions that are not favorable to germination, very often high temperatures. The mechanisms of secondary dormancy are not yet fully understood, but might involve the loss of sensitivity in receptors in the plasma membrane.

The following types of seed dormancy do not involve seed dormancy, strictly speaking, as lack of germination is prevented by the environment, not by characteristics of the seed itself (see Germination):

- **Photodormancy** or light sensitivity affects germination of some seeds. These photoblastic seeds need a period of darkness or light to germinate. In species with thin seed coats, light may be able to penetrate into the dormant embryo. The presence of light or the absence of light may trigger the germination process, inhibiting germination in some seeds buried too deeply or in others not buried in the soil.
- **Thermodormancy** is seed sensitivity to heat or cold. Some seeds, including cocklebur and amaranth, germinate only at high temperatures (30 °C or 86 °F); many plants that have seeds that germinate in early to midsummer have thermodormancy, so germinate only when the soil temperature is warm. Other seeds need cool soils to germinate, while others, such as celery, are inhibited when soil temperatures are too warm. Often, thermodormancy requirements disappear as the seed ages or dries.

Not all seeds undergo a period of dormancy. Seeds of some mangroves are viviparous; they begin to germinate while still attached to the parent. The large, heavy root allows the seed to penetrate into the ground when it falls. Many garden plant seeds will germinate readily as soon as they have water and are warm enough; though their wild ancestors may have had dormancy, these cultivated plants lack it. After many generations of selective pressure by plant breeders and gardeners, dormancy has been selected out.

For annuals, seeds are a way for the species to survive dry or cold seasons. Ephemeral plants are usually annuals that can go from seed to seed in as few as six weeks.

Persistence and Seed Banks

Germination

Seed germination is a process by which a seed embryo develops into a seedling. It involves the reactivation of the metabolic pathways that lead to growth and the emergence of the radicle or seed root and plumule or shoot. The emergence of the seedling above the soil surface is the next phase of the plant's growth and is called seedling establishment.

Three fundamental conditions must exist before germination can occur. (1) The embryo must be alive, called seed viability. (2) Any dormancy requirements that prevent germination must be overcome. (3) The proper environmental conditions must exist for germination.

Seed viability is the ability of the embryo to germinate and is affected by a number of different conditions. Some plants do not produce seeds that have functional complete embryos, or the seed may have no embryo at all, often called empty seeds. Predators and pathogens can damage or kill the seed while it is still in the fruit or after it is dispersed. Environmental conditions like flooding or heat can kill the seed before or during germination. The age of the seed affects its health and germination ability: since the seed has a living embryo, over time cells die and cannot be replaced. Some seeds can live for a long time before germination, while others can only survive for a short period after dispersal before they die.

Germinating sunflower seedlings

Seed vigor is a measure of the quality of seed, and involves the viability of the seed, the germination percentage, germination rate and the strength of the seedlings produced.

The **germination percentage** is simply the proportion of seeds that germinate from all seeds subject to the right conditions for growth. The **germination rate** is the length of time it takes for the seeds to germinate. Germination percentages and rates are affected by seed viability, dormancy and environmental effects that impact on the seed and seedling. In agriculture and horticulture quality seeds have high viability, measured by germination percentage plus the rate of germination. This is given as a percent of germination over a certain amount of time, 90% germination in 20 days, for example. 'Dormancy' is covered above; many plants produce seeds with varying degrees of dormancy, and different seeds from the same fruit can have different degrees of dormancy. It's possible to have seeds with no dormancy if they are dispersed right away and do not dry (if the seeds dry they go into physiological dormancy). There is great variation amongst plants and a dormant seed is still a viable seed even though the germination rate might be very low.

Environmental conditions affecting seed germination include; water, oxygen, temperature and light.

Three distinct phases of seed germination occur: water imbibition; lag phase; and radicle emergence.

In order for the seed coat to split, the embryo must imbibe (soak up water), which causes it to swell, splitting the seed coat. However, the nature of the seed coat determines how rapidly water can penetrate and subsequently initiate germination. The rate of imbibition is dependent on the permeability of the seed coat, amount of water in the environment and the area of contact the seed has to the source of water. For some seeds, imbibing too much water too quickly can kill the seed. For some seeds, once water is imbibed the germination process cannot be stopped, and drying then becomes fatal. Other seeds can imbibe and lose water a few times without causing ill effects, but drying can cause secondary dormancy.

Repair of DNA Damage

During seed dormancy, often associated with unpredictable and stressful environments, DNA damages accumulate as the seeds age. In rye seeds, the reduction of DNA integrity due to damage is associated with loss of seed viability during storage. Upon germination, seeds of *Vicia faba* undergo DNA repair. A plant DNA ligase that is involved in repair of single- and double-strand breaks during seed germination is an important determinant of seed longevity. Also, in Arabidopsis seeds, the activities of the DNA repair enzymes Poly ADP ribose polymerases (PARP) are likely needed for successful germination. Thus DNA damages that accumulate during dormancy appear to be a problem for seed survival, and the enzymatic repair of DNA damages during germination appears to be important for seed viability.

Inducing Germination

A number of different strategies are used by gardeners and horticulturists to break seed dormancy.

Scarification allows water and gases to penetrate into the seed; it includes methods to physically break the hard seed coats or soften them by chemicals, such as soaking in hot water or poking holes in the seed with a pin or rubbing them on sandpaper or cracking with a press or hammer. Sometimes fruits are harvested while the seeds are still immature and the seed coat is not fully developed and sown right away before the seed coat become impermeable. Under natural conditions, seed coats are worn down by rodents chewing on the seed, the seeds rubbing against rocks (seeds are moved by the wind or water currents), by undergoing freezing and thawing of surface water, or passing through an animal's digestive tract. In the latter case, the seed coat protects the seed from digestion, while often weakening the seed coat such that the embryo is ready to sprout when it is deposited, along with a bit of fecal matter that acts as fertilizer, far from the parent plant. Microorganisms are often effective in breaking down hard seed coats and are sometimes used by people as a treatment; the seeds are stored in a moist warm sandy medium for several months under nonsterile conditions.

Stratification, also called moist-chilling, breaks down physiological dormancy, and involves the addition of moisture to the seeds so they absorb water, and they are then subjected to a period of moist chilling to after-ripen the embryo. Sowing in late summer and fall and allowing to overwinter under cool conditions is an effective way to stratify seeds; some seeds respond more favorably to periods of oscillating temperatures which are a part of the natural environment.

Leaching or the soaking in water removes chemical inhibitors in some seeds that prevent germination. Rain and melting snow naturally accomplish this task. For seeds planted in gardens, running water is best—if soaked in a container, 12 to 24 hours of soaking is sufficient. Soaking longer, especially in stagnant water, can result in oxygen starvation and seed death. Seeds with hard seed coats can be soaked in hot water to break open the impermeable cell layers that prevent water intake.

Other methods used to assist in the germination of seeds that have dormancy include prechilling, predrying, daily alternation of temperature, light exposure, potassium nitrate, the use of plant growth regulators, such as gibberellins, cytokinins, ethylene, thiourea, sodium hypochlorite, and others. Some seeds germinate best after a fire. For some seeds, fire cracks hard seed coats, while in others, chemical dormancy is broken in reaction to the presence of smoke. Liquid smoke is often used by gardeners to assist in the germination of these species.

Sterile Seeds

Seeds may be sterile for few reasons: they may have been irradiated, unpollinated, cells lived past expectancy, or bred for the purpose.

Evolution and Origin of Seeds

The origin of seed plants is a problem that still remains unsolved. However, more and more data tends to place this origin in the middle Devonian. The description in 2004 of the proto-seed *Runcaria heinzelinii* in the Givetian of Belgium is an indication of that ancient origin of seed-plants. As with modern ferns, most land plants before this time reproduced by sending spores into the air, that would land and become whole new plants.

The first "true" seeds are described from the upper Devonian, which is probably the theater of their true first evolutionary radiation. The seed plants progressively became one of the major elements of nearly all ecosystems.

Economic Importance

Phaseolus vulgaris seeds are diverse in size, shape, and color.

Edible Seeds

Many seeds are edible and the majority of human calories comes from seeds, especially from cereals, legumes and nuts. Seeds also provide most cooking oils, many beverages and spices and some important food additives. In different seeds the seed embryo or the endosperm dominates and provides most of the nutrients. The storage proteins of the embryo and endosperm differ in their amino acid content and physical properties. For example, the gluten of wheat, important in providing the elastic property to bread dough is strictly an endosperm protein.

Seeds are used to propagate many crops such as cereals, legumes, forest trees, turfgrasses and pasture grasses. Particularly in developing countries, a major constraint faced is the inadequacy of the marketing channels to get the seed to poor farmers. Thus the use of farmer-retained seed remains quite common.

Seeds are also eaten by animals, and are fed to livestock. Many seeds are used as birdseed.

Poison and Food Safety

While some seeds are edible, others are harmful, poisonous or deadly. Plants and seeds often contain chemical compounds to discourage herbivores and seed predators. In some cases, these compounds simply taste bad (such as in mustard), but other compounds are toxic or break down into toxic compounds within the digestive system. Children, being smaller than adults, are more susceptible to poisoning by plants and seeds.

A deadly poison, ricin, comes from seeds of the castor bean. Reported lethal doses are anywhere from two to eight seeds, though only a few deaths have been reported when castor beans have been ingested by animals.

In addition, seeds containing amygdalin—apple, apricot, bitter almond, peach, plum, cherry, quince, and others—when consumed in sufficient amounts, may cause cyanide poisoning. Other seeds that contain poisons include annona, cotton, custard apple, datura, uncooked durian, golden chain, horse-chestnut, larkspur, locoweed, lychee, nectarine, rambutan, rosary pea, sour sop, sugar apple, wisteria, and yew. The seeds of the strychnine tree are also poisonous, containing the poison strychnine.

The seeds of many legumes, including the common bean (*Phaseolus vulgaris*), contain proteins called lectins which can cause gastric distress if the beans are eaten without cooking. The common bean and many others, including the soybean, also contain trypsin inhibitors which interfere with the action of the digestive enzyme trypsin. Normal cooking processes degrade lectins and trypsin inhibitors to harmless forms.

Please see the category plant toxins for further relevant articles.

Other Uses

Cotton fiber grows attached to cotton plant seeds. Other seed fibers are from kapok and milkweed.

Many important nonfood oils are extracted from seeds. Linseed oil is used in paints. Oil from jojoba and crambe are similar to whale oil.

Seeds are the source of some medicines including castor oil, tea tree oil and the cancer drug, Laetrile.

Many seeds have been used as beads in necklaces and rosaries including Job`s tears, Chinaberry, rosary pea, and castor bean. However, the latter three are also poisonous.

Other seed uses include:

- Seeds once used as weights for balances.
- Seeds used as toys by children, such as for the game Conkers.
- Resin from *Clusia rosea* seeds used to caulk boats.
- Nematicide from milkweed seeds.
- Cottonseed meal used as animal feed and fertilizer.

Seed Records

The massive fruit of the coco de mer.

- The oldest viable carbon-14-dated seed that has grown into a plant was a Judean date palm seed about 2,000 years old, recovered from excavations at Herod the Great's palace on Masada in Israel. It was germinated in 2005. (A reported regeneration of *Silene stenophylla* (narrow-leafed campion) from material preserved for 31,800 years in the Siberianpermafrost was achieved using fruit tissue, not seed.)

- The largest seed is produced by the coco de mer, or "double coconut palm", *Lodoicea maldivica*. The entire fruit may weigh up to 23 kilograms (50 pounds) and usually contains a single seed.
- The smallest seeds are produced by epiphytic orchids. They are only 85 micrometers long, and weigh 0.81 micrograms. They have no endosperm and contain underdeveloped embryos.
- The earliest fossil seeds are around 365 million years old from the Late Devonian of West Virginia. The seeds are preserved immature ovules of the plant *Elkinsia polymorpha*.

In Religion

The Book of Genesis in the Old Testament begins with an explanation of how all plant forms began:

And God said, Let the earth bring forth grass, the herb yielding seed, and the fruit tree yielding fruit after his kind, whose seed is in itself, upon the earth: and it was so. And the earth brought forth grass, and herb yielding seed after its kind, and the tree yielding fruit, whose seed was in itself, after its kind: and God saw that it was good. And the evening and the morning were the third day.

The Quran speaks about seed germination:

It is Allah Who causeth the seed-grain and the date-stone to split and sprout. He causeth the living to issue from the dead, and He is the one to cause the dead to issue from the living. That is Allah: then how are ye deluded away from the truth?

REFERENCES

1. Harms, K; Wright, SJ; Calderon, O; Hernandez, A; Herre, EA (2000). "Pervasive density-dependent recruitment enhances seedling diversity in a tropical forest". Nature. 404 (6777): 493–495. doi:10.1038/35006630. PMID 10761916.
2. Wenny, D.G. & Levey, D.J. (1998). "Directed seed dispersal by bellbirds in a tropical cloud forest". Proceedings of the National Academy of Sciences of the United States of America. 95 (11): 6204–7. doi:10.1073/pnas.95.11.6204. PMC 27627 . PMID 9600942.
3. Fedriani, J. M., Delibes, M. (2009). "Functional diversity in fruit-frugivore interactions: A field experiment with Mediterranean mammals". Ecography. 32 (6): 983. doi:10.1111/j.1600-0587.2009.05925.x.
4. Lengyel, S.; et al. (2010). "Convergent evolution of seed dispersal by ants, and phylogeny and biogeography in flowering plants: a global survey". Perspectives in Plant Ecology, Evolution and Systematics. 12 (1): 43–55. doi:10.1016/j.ppees.2009.08.001.
5. Manzaneda, Antonio J.; Fedriani, Jose M. & Rey, Pedro J. (2005). "Adaptive advantages of myrmecochory: the predator-avoidance hypothesis tested over a wide geographic range" (PDF). Ecography. 28 (5): 583–592. doi:10.1111/

j.2005.0906-7590.04309.x.

6. Manzano, Pablo; Malo, Juan E. (2006). "Extreme long-distance seed dispersal via sheep". Frontiers in Ecology and the Environment. 4 (5): 244–248. doi:10.1890/1540-9295(2006)004 0244:ELSDVS2.0.CO;2. JSTOR 3868790.
7. Vittoz, Pascal; Engler, Robin (7 February 2008). "Seed dispersal distances: a typology based on dispersal modes and plant traits" (PDF). Botanica Helvetica. 117 (2): 109–124. doi:10.1007/s00035-007-0797-8. Retrieved 23 June 2016.
8. Schulze, Ernst-Detlef; Beck, Erwin & Müller-Hohenstein, Klaus (2005). Plant Ecology. Springer. pp. 543–. ISBN 978-3-540-20833-4.
9. "Dispersal of seeds by gravity". Retrieved 2009-05-08.
10. Feldkamp, Susan (2006). Modern Biology. United States: Holt, Rinehart, and Winston. p. 618.
11. Gurevitch, J., Scheiner, S.M., & G.A. Fox (2006). Plant Ecology, 2nd ed. Sinauer Associates, Inc., Massachusetts.
12. Cody, M.L. & Overton, J.M. (1996). "Short-term evolution of reduced dispersal in island plant populations". Journal of Ecology. 84: 53–61. doi:10.2307/2261699. JSTOR 2261699.
13. Godt, Mary (June 1995). "Genetic Diversity in a Threatened Wetland Species, Helonias bullata (Liliaceae)". Conservation Biology. Wiley for Society for Conservation Biology. 9 (3): 596–604. doi:10.1046/j.1523-1739.1995.09030596.x. JSTOR 2386613.
14. Sorenson, A.E. (1986). "Seed dispersal by adhesion". Annual Review of Ecology and Systematics. 17: 443–463. doi:10.1146/annurev.es.17.110186.002303.
15. Howe, H. F. & Smallwood J. (1982). "Ecology of Seed Dispersal" (PDF). Annual Review of Ecology and Systematics. 13: 201–228. doi:10.1146/annurev.es.13.110182.001221.
16. Corlett, R.T. (1998). "Frugivory and seed dispersal by vertebrates in the Oriental (Indomalayan) Region". Biological Reviews. 73 (4): 413–448. doi:10.1017/S0006323198005234. PMID 9951414.
17. Terborgh, J. (1986) "Community aspects of frugivory in tropical forests": in Fleming, T.H.; Estrada, Alejandro (eds.) Frugivory and Seed Dispersal, Advances in Vegetation Science, Vol. 15, Springer, ISBN 978-0-7923-2141-5.
18. Chapman, C.A. & Onderdonk, D.A. (1998). "Forests without primates: primate/plant codependency". American Journal of Primatology. 45 (1): 127–141. doi:10.1002/(SICI)1098-2345(1998)45:1<127::AID-AJP9>3.0.CO;2-Y. PMID 9573446.
19. Giladi, I. (2006). "Choosing benefits or partners: a review of the evidence for the evolution of myrmecochory". Oikos. 112 (3): 481–492. doi:10.1111/j.0030-1299.2006.14258.x.
20. Forget, P.M. & Milleron, T. (1991). "Evidence for secondary seed dispersal by rodents in Panama". Oecologia. 87 (4): 596–599. doi:10.1007/BF00320426.
21. Samuni-Blank, M.; et al. (2012). "Intraspecific directed deterrence by the mustard

oil bomb in a desert plant". Current Biology. 22 (13): 1–3. doi:10.1016/j.cub.2012.04.051. PMID 22704992.

22. Andresen E. & Levey, D.J. (2004). "Effects of dung and seed size on secondary dispersal, seed predation, and seedling establishment of rainforest trees". Oecologia. 139 (1): 45–54. doi:10.1007/s00442-003-1480-4. PMID 14740290.
23. Wichmann, M.C.; Alexander, M.J.; Soons, M.B.; Galsworthy, S.; Dunne, L.; Gould, R.; Fairfax, C.; Niggemann, M.; Hails, R.S. & Bullock, J.M. (2009). "Human mediated dispersal of seeds over long-distances". Proceedings of the Royal Society B. 276 (1656): 523–532. doi:10.1098/rspb.2008.1131.
24. Chaloupka, M. Y.; Domm, S. B. (Dec 1986). "Role of Anthropochory in the Invasion of Coral Cays by Alien Flora". Ecology. Ecological Society of America. 67 (6): 1536–1547. doi:10.2307/1939084. JSTOR 1939084.
25. "Anthropochory or Human-Mediated Dispersal (HMD)". Frugivores and Seed Dispersal Symposium. June 2010.
26. Bullock, S. H. & Primack, R. B. (1977). "Comparative experimental study of seed dispersal on animals". Ecology. 58 (3): 681–686. doi:10.2307/1939019.
27. von der Lippe, M. & Kowarik, I. (2007). "Long-distance dispersal of plants by vehicles as a driver of plant invasions". Conservation Biology. 21 (4): 986–996. doi:10.1111/j.1523-1739.2007.00722.x. PMID 17650249.
28. Caswell, H.; Lensink, R.; Neubert, M.G. (2003). "Demography And Dispersal: Life Table Response Experiments For Invasion Speed". Ecology. 84 (8): 1968–1978. doi:10.1890/02-0100.
29. Lengyel, S.; et al. (2009). Chave, Jerome, ed. "Ants Sow the Seeds of Global Diversification in Flowering Plants". PLoS ONE. 4 (5): e5480. doi:10.1371/journal.pone.0005480. PMC 2674952 . PMID 19436714.
30. Cain, M.D.; Shelton, M.G. 2001. Twenty years of natural loblolly and shortleaf pine seed production on the Crossett Experimental Forest in southeastern Arkansas. Southern Journal of Applied Forestry. 25(1): 40-45.
31. Galili G; Kigel J (1995). "Chapter One". Seed development and germination. New York: M. Dekker. ISBN 0-8247-9229-7.
32. Raven, Peter H., Ray Franklin Evert, and Helena Curtis. 1981. Biology of plants. New York, N.Y.: Worth Publishers. page 410.
33. Rost, Thomas L.; Weier, T. Elliot; Weier, Thomas Elliot (1979). Botany: a brief introduction to plant biology. New York: Wiley. p. 319. ISBN 0-471-02114-8.
34. Filonova LH; Bozhkov PV; von Arnold S (February 2000). "Developmental pathway of somatic embryogenesis in Picea abies as revealed by time-lapse tracking". J Exp Bot. 51 (343): 249–64. doi:10.1093/jexbot/51.343.249. PMID 10938831.
35. "Seed shape". anbg.gov.au.
36. The Seed Biology Place, Gerhard Leubner Lab, Royal Holloway, University of London, retrieved 13 October 2015
37. "Carol C. Baskin, Jerry M. Baskin. Seeds: Ecology, Biogeography, and Evolution

of Dormancy and Germination. Elsevier, 2001". google.ca. p. 27. ISBN 0-12-080263-5.

38. "The Encyclopaedia Britannica, 9th ed. (1888) vol. 4,". google.ca. p. 155.
39. Bewley & Black (1978) Physiology and Biochemistry of Seeds in Relation to Germination, pag.11
40. Plant Physiology online: Types of Seed Dormancy and the Roles of Environmental Factors
41. "plant_anatomy Term "seed coat epidermis" (PO:0006048)". gramene.org.
42. "6 – Seed and fruit – University Publishing Online – Paula J. Rudall. Anatomy of Flowering Plants: An Introduction to Structure and Development. Third edition". Cambridge University Press. 2007. ISBN 978-0-521-69245-8.
43. Smith, Welby R. 1993. Orchids of Minnesota. Minneapolis: University of Minnesota Press. Page 8.
44. Igor Kosinki (2007). "Long-term variability in seed size and seedling establishment of Maianthemum bifolium". Plant Ecology. 194 (2): 149–156. doi:10.1007/s11258-007-9281-1.
45. Shannon DA; Isaac L; Brockman FE (February 1996). "Assessment of hedgerow species for seed size, stand establishment and seedling height". Agroforestry Systems. 35 (1): 95–110. doi:10.1007/BF02345331.
46. Jones, Samuel B., and Arlene E. Luchsinger. 1979. Plant systematics. McGraw-Hill series in organismic biology. New York: McGraw-Hill. Page 195.
47. Morhardt, Sia; Morhardt, Emil; Emil Morhardt, J. (2004). California desert flowers: an introduction to families, genera, and species. Berkeley: University of California Press. p. 24. ISBN 0-520-24003-0.
48. "www.seabean.com – Sea-Beans and Drift Seeds". seabean.com.
49. Marinelli, J. (1999) "Ants – The astonishing intimacy between ants & plants." Plants & Gardens News 14 (1).
50. Ricklefs, Robert E. (1993) The Economy of Nature, 3rd ed., p.396. (New York: W. H. Freeman). ISBN 0-7167-2409-X.
51. Bond, W. J.; P. Slingsby (1984). "Collapse of an ant-plant mutualism: The Argentine ant, Iridomyrmex humilis and myrmecochorous Proteaceae". Ecology. 65 (4): 1031–1037. doi:10.2307/1938311. JSTOR 1938311.
52. Eira MTS, Caldas LS (2000) Seed dormancy and germination as concurrent processes. Rev Bras Fisiol Vegetal 12:85–104
53. Vleeshouwers, L.M., H.J. Bouwmeester & C.M. Karssen, 1995. Redefining seed dormancy: an attempt to integrate physiology and ecology. Journal of Ecology 83:1031-1037
54. Thompson K, Ceriani RM, Bakker JP, Bekker RM (2003) Are seed dormancy and persistence in soil related? 13:97-100
55. Baskin, J.M. and Baskin, C.C. (2004) A classification system for seed dormancy. Seed Science Research 14, 1–16

56. Baskin JM, Baskin CC, Li X (2000) Taxonomy, anatomy and evolution of physical dormancy in seeds. Plant Species Biology 15:139-152
57. Baskin, C.C. and Baskin, J.M. (1998) Seeds: Ecology, biogeography, and evolution of dormancy and germination.San Diego, Academic Press
58. Gutterman, Y. (1993) Seed germination in desert plants. Springer Verlag, Berlin/ Heidelberg.
59. Baskin, C.C. and Baskin, J.M. (1998) Seeds: Ecology, biogeography, and evolution of dormancy and germination.San Diego, Academic Press.
60. Baskin, J.M. and Baskin, C.C. (2004) A classification system for seed dormancy. Seed Science Research 14:1–16.
61. International Workshop on Seeds, and G. Nicolas. 2003. The biology of seeds recent research advances : proceedings of the Seventh International Workshop on Seeds, Salamanca, Spain 2002. Wallingford, Oxon, UK: CABI Pub. Page 113.
62. Bewley, J. Derek, and Michael Black. 1994. Seeds physiology of development and germination. The language of science. New York: Plenum Press. page 230.
63. Patten D.T. (1978). "Productivity and production efficiency of an Upper Sonoran Desert ephemeral community". American Journal of Botany. 65 (8): 891–895. doi:10.2307/2442185.
64. Black, Michael H.; Halmer, Peter (2006). The encyclopedia of seeds: science, technology and uses. Wallingford, UK: CABI. p. 224. ISBN 978-0-85199-723-0.
65. Seed Vigor and Vigor Tests
66. International Seed Testing Association. 1973. ISSN 0251-0952. Pages 120-21.Seed science and technology. Wageningen?: International Seed Testing Association.
67. Cheah KS; Osborne DJ (April 1978). "DNA lesions occur with loss of viability in embryos of ageing rye seed". Nature. 272 (5654): 593–9. doi:10.1038/272593a0. PMID 19213149.
68. Koppen G; Verschaeve L (2001). "The alkaline single-cell gel electrophoresis/ comet assay: a way to study DNA repair in radicle cells of germinating Vicia faba". Folia Biol. (Praha). 47 (2): 50–4. PMID 11321247.
69. Bray CM; West CE (December 2005). "DNA repair mechanisms in plants: crucial sensors and effectors for the maintenance of genome integrity". New Phytol. 168 (3): 511–28. doi:10.1111/j.1469-8137.2005.01548.x. PMID 16313635.
70. Waterworth WM; Masnavi G; Bhardwaj RM; Jiang Q; Bray CM; West CE (September 2010). "A plant DNA ligase is an important determinant of seed longevity". Plant J. 63 (5): 848–60. doi:10.1111/j.1365-313X.2010.04285.x. PMID 20584150.
71. Hunt L; Holdsworth MJ; Gray JE (August 2007). "Nicotinamidase activity is important for germination". Plant J. 51 (3): 341–51. doi:10.1111/j.1365-313X.2007.03151.x. PMID 17587307.
72. Hartmann, Hudson Thomas, and Dale E. Kester. 1983. Plant propagation principles and practices. Englewood Cliffs, N.J.: Prentice-Hall. ISBN 0-13-681007-1. Pages 175-77.

73. Jon E. Keeley and Fotheringham. "Trace Gas Emissions and Smoke-Induced Seed Germination". Science. p. 1248.
74. Sabelli, P.A.; Larkins, B.A. (2009). "The Development of Endosperm in Grasses". Plant Physiology. 149 (1): 14–26. doi:10.1104/pp.108.129437. PMC 2613697 . PMID 19126691.
75. G. Mumby Seed Marketing, FAO, Rome
76. Chia Joo Suan, "Seeds of Doubt: Food Safety"
77. Clelland, Mike. "Poisonous Plants and Seeds", Healthy Child Care
78. Martin Anderson, Texas AgriLife Extension Service. "Poisonous Plants and Plant Parts – Archives – Aggie Horticulture". tamu.edu.
79. Wedin GP; Neal JS; Everson GW; Krenzelok EP (May 1986). "Castor bean poisoning". Am J Emerg Med. 4 (3): 259–61. doi:10.1016/0735-6757(86)90080-X. PMID 3964368.
80. Albretsen JC; Gwaltney-Brant SM; Khan SA (2000). "Evaluation of castor bean toxicosis in dogs: 98 cases". J Am Anim Hosp Assoc. 36 (3): 229–33. PMID 10825094.
81. "Almond/Almond Oil". drugs.com.
82. Wolke, RL. Seeds of Anxiety Washington Post January 5, 2005
83. Chia Joo Suan Food Safety: Seeds of doubt
84. Dhurandhar NV; Chang KC (1990). "Effect of Cooking on Firmness, Trypsin Inhibitors, Lectins and Cystine/Cysteine content of Navy and Red Kidney Beans (Phaseolus vulgaris)". J Food Sci. 55 (2): 470–4. doi:10.1111/j.1365-2621.1990.tb06789.x.
85. Roach, John. (2005) "2,000-Year-Old Seed Sprouts, Sapling Is Thriving", National Geographic News, 22 November.
86. "Ice Age flower revival that could lead to resurrection of mammoth". Telegraph.co.uk. 21 February 2012.
87. "Russian Scientists Revive 32,000-Year-Old Flower". sci-news.com.
88. Corner EJH (1966). The Natural History of Palms. Berkeley, CA: University of California Press. pp. 313–4.
89. http://waynesword.palomar.edu/ww0601.htm#seed
90. Taylor EL; Taylor TMC (1993). The biology and evolution of fossil plants. Englewood Cliffs, N.J: Prentice Hall. p. 466. ISBN 0-13-651589-4.
91. King James Version, Genesis 1:12,13, 1611.
92. Quran, Translation: Abdullah Yusuf Ali, Al-An'aam 95:6

CHAPTER 6

Tillage Systems and Techniques

Chapter Contents

Soil tillage is a method of soil preparation for seedbed preparation, sowing or transplanting, and for crops' growth.

Types of soil tillage

Conventional tillage

The cultivation of the soil using plow, harrow and other farm tools or mechanical implements to prepare the field for crop production.

Advantages

- Destroys pests' shelters and disrupts their lifecycles
- Exposes pests to predators and unfavorable conditions
- Distributes soil nutrients throughout the soil
- Aerates the soil
- Controls weeds
- Makes other farm cultural practices easier to undertake

INTRODUCTION

Tillage is the agricultural preparation of soil by mechanical agitation of various types, such as digging, stirring, and overturning. Examples of human-powered **tilling** methods using hand tools include shovelling, picking, mattock work, hoeing, and raking. Examples of draft-animal-powered or mechanized work include ploughing (overturning with moldboards or chiseling with chisel shanks), rototilling, rolling with cultipackers or other rollers, harrowing, and cultivating with cultivator shanks (teeth). Small-scale gardening and farming, for household food production or small business production, tends to use the smaller-scale methods above, whereas medium- to large-scale farming tends to use the larger-scale methods. There is a fluid continuum, however. Any type of gardening or farming, but especially larger-scale commercial types, may also use low-till or no-till methods as well.

Cultivating after an early rain.

Tillage is often classified into two types, primary and secondary. There is no strict boundary between them so much as a loose distinction between tillage that is deeper and more thorough (primary) and tillage that is shallower and sometimes more selective of location (secondary). Primary tillage such as ploughing tends to produce a rough surface finish, whereas secondary tillage tends to produce a smoother surface finish, such as that required to make a good seedbed for many crops. Harrowing and rototilling often combine primary and secondary tillage into one operation.

"Tillage" can also mean the land that is **tilled**. The word "**cultivation**" has several senses that overlap substantially with those of "tillage". In a general context, both can refer to agriculture. Within agriculture, both can refer to any of the kinds of soil agitation described above. Additionally, "cultivation" or "cultivating" may refer to an even narrower sense of shallow, selective secondary tillage of row crop fields that kills weeds while sparing the crop plants.

TILLAGE SYSTEMS

Reduced Tillage

Reduced tillage leaves between 15 and 30% residue cover on the soil or 500 to 1000 pounds per acre (560 to 1100 kg/ha) of small grain residue during the critical erosion period. This may involve the use of a chisel plow, field cultivators, or other implements. See the general comments below to see how they can affect the amount of residue.

Plough tilling the field

Intensive Tillage

Intensive tillage leaves less than 15% crop residue cover or less than 500 pounds per acre (560 kg/ha) of small grain residue. This type of tillage is often referred to as conventional tillage but as conservational tillage is now more widely used than intensive tillage (in the United States), it is often not appropriate to refer to this type of tillage as conventional. Intensive tillage often involves multiple operations with implements such as a mold board, disk, and/or chisel plow. Then a finisher with a harrow, rolling basket, and cutter can be used to prepare the seed bed. There are many variations.

Conservation Tillage

Conservation tillage leaves at least 30% of crop residue on the soil surface, or at least 1,000 lb/ac (1,100 kg/ha) of small grain residue on the surface during the critical soil erosion period. This slows water movement, which reduces the amount of soil erosion. Conservation tillage also benefits farmers by reducing fuel consumption and soil compaction.

By reducing the number of times the farmer travels over the field, farmers realize significant savings in fuel and labor. In most years since 1997, conservation tillage was used in US cropland more than intensive or reduced tillage.

However, conservation tillage delays warming of the soil due to the reduction of dark earth exposure to the warmth of the spring sun, thus delaying the planting of the next year's spring crop of corn.

- No-till - Never use a plow, disk, etc. ever again. Aims for 100% ground cover.
- Strip-Till - Narrow strips are tilled where seeds will be planted, leaving the soil in between the rows untilled.
- Mulch-till
- Rotational Tillage - Tilling the soil every two years or less often (every other year, or every third year, etc.).
- Ridge-Till
- Zone tillage - This form of conservation tillage is further explained below.

ZONE TILLAGE

Zone tillage is a form of modified deep tillage in which only narrow strips are tilled, leaving soil in between the rows untilled. This type of tillage agitates the soil to help reduce soil compaction problems and to improve internal soil drainage.

Purpose

Zone tillage is designed to only disrupt the soil in a narrow strip directly below the crop row. In comparison to no-till, which relies on the previous year's plant residue to protect the soil and aides in postponement of the warming of the soil and crop growth in Northern climates, zone tillage creates approximately a 5-inch-wide strip that simultaneously breaks up plow pans, assists in warming the soil and helps to prepare a seedbed. When combined with cover crops, zone tillage helps replace lost organic matter, slows the deterioration of the soil, improves soil drainage, increases soil water and nutrient holding capacity, and allows necessary soil organisms to survive.

Usage

It has been successfully used on farms in the mid-west and west for over 40 years and is currently used on more than 36% of the U.S. farmland. Some specific states where zone tillage is currently in practice are Pennsylvania, Connecticut, Minnesota, Indiana, Wisconsin, and Illinois.

Unfortunately, there aren't consistent yield results in the Northern Cornbelt states; however, there is still interest in deep tillage within the agriculture industry. In areas that are not well-drained, deep tillage may be used as an alternative to installing more expensive tile drainage.

Effects of Tillage

Positive

Plowing:

- Loosens and aerates the top layer of soil, which facilitates planting the crop
- Helps mix harvest residue, organic matter (humus), and nutrients evenly into the soil
- Mechanically destroys weeds
- Dries the soil before seeding (in wetter climates tillage aids in keeping the soil drier)
- When done in autumn, helps exposed soil crumble over winter through frosting and defrosting, which helps prepare a smooth surface for spring planting

Negative

- Dries the soil before seeding
- Soil loses a lot of nutrients, like nitrogen and fertilizer, and its ability to store water
- Decreases the water infiltration rate of soil. (Results in more runoff and erosion since the soil absorbs water slower than before)
- Tilling the soil results in dislodging the cohesiveness of the soil particles thereby inducing erosion.
- Chemical runoff
- Reduces organic matter in the soil
- Reduces microbes, earthworms, ants, etc.
- Destroys soil aggregates
- Compaction of the soil, also known as a tillage pan
- Eutrophication (nutrient runoff into a body of water)
- Can attract slugs, cut worms, army worms, and harmful insects to the left over residues.
- Crop diseases can be harbored in surface residues

General Comments

- The type of implement makes the most difference, although other factors can have an effect.
- Tilling in absolute darkness (night tillage) might reduce the number of weeds that sprout following the tilling operation by half. Light is necessary to break the dormancy of some weed species' seed, so if fewer seeds are exposed to

light during the tilling process, fewer will sprout. This may help reduce the amount of herbicides needed for weed control.

- Greater speeds, when using certain tillage implements (disks and chisel plows), lead to more intensive tillage (i.e., less residue is on the soil surface).
- Increasing the angle of disks causes residues to be buried more deeply. Increasing their concavity makes them more aggressive.
- Chisel plows can have spikes or sweeps. Spikes are more aggressive.
- Percentage residue is used to compare tillage systems because the amount of crop residue affects the soil loss due to erosion.
- See Soybean management practices to see what types of tillage are currently recommended for Soybean Production.

Primary tillage loosens the soil and mixes in fertilizer and/or plant material, resulting in soil with a rough texture.

Secondary tillage produces finer soil and sometimes shapes the rows, preparing the seed bed. It also provides weed control throughout the growing season during the maturation of the crop plants, unless such weed control is instead achieved with low-till or no-till methods involving herbicides.

- The seed bed preparation can be done with harrows (of which there are many types and subtypes), dibbles, hoes, shovels, rotary tillers, subsoilers, ridge- or bed-forming tillers, rollers, or cultivators.
- The weed control, to the extent that it is done via tillage, is usually achieved with cultivators or hoes, which disturb the top few centimeters of soil around the crop plants but with minimal disturbance of the crop plants themselves. The tillage kills the weeds via 2 mechanisms: uprooting them, burying their leaves (cutting off their photosynthesis), or a combination of both. Weed control both prevents the crop plants from being outcompeted by the weeds (for water and sunlight) and prevents the weeds from reaching their seed stage, thus reducing future weed population aggressiveness.

History of Tilling

Tilling was first performed via human labor, sometimes involving slaves. Hoofed animals could also be used to till soil via trampling. The wooden plow was then invented. It could be pulled by mule, ox, elephant, water buffalo, or similar sturdy animal. Horses are generally unsuitable, though breeds such as the scyne could work. The steel plow allowed farming in the American Midwest, where tough prairie grasses and rocks caused trouble. Soon after 1900, the farm tractor was introduced, which eventually made modern large-scale agriculture possible.

Tilling with Hungarian Grey cattles.

Alternatives to tilling

Modern agricultural science has greatly reduced the use of tillage. Crops can be grown for several years without any tillage through the use of herbicides to control weeds, crop varieties that tolerate packed soil, and equipment that can plant seeds or fumigate the soil without really digging it up. This practice, called no-till farming, reduces costs and environmental change by reducing soil erosion and diesel fuel usage.

Site preparation of forest land

Site preparation is any of various treatments applied to a site in order to ready it for seeding or planting. The purpose is to facilitate the regeneration of that site by the chosen method. Site preparation may be designed to achieve, singly or in any combination: improved access, by reducing or rearranging slash, and amelioration of adverse forest floor, soil, vegetation, or other biotic factors. Site preparation is undertaken to ameliorate one or more constraints that would otherwise be likely to thwart the objectives of management. A valuable bibliography on the effects of soil temperature and site preparation on subalpine and boreal tree species has been prepared by McKinnon et al. (2002).

Site preparation is the work that is done before a forest area is regenerated. Some types of site preparation are burning.

Burning

Broadcast burning is commonly used to prepare clearcut sites for planting, e.g., in central British Columbia, and in the temperate region of North America generally.

Prescribed burning is carried out primarily for slash hazard reduction and to improve site conditions for regeneration; all or some of the following benefits may accrue:

- Reduction of logging slash, plant competition, and humus prior to direct seeding, planting, scarifying or in anticipation of natural seeding in partially cut stands or in connection with seed-tree systems.
- Reduction or elimination of unwanted forest cover prior to planting or seeding, or prior to preliminary scarification thereto.
- Reduction of humus on cold, moist sites to favour regeneration.
- Reduction or elimination of slash, grass, or brush fuels from strategic areas around forested land to reduce the chances of damage by wildfire.

Prescribed burning for preparing sites for direct seeding was tried on a few occasions in Ontario, but none of the burns was hot enough to produce a seedbed that was adequate without supplementary mechanical site preparation.

Changes in soil chemical properties associated with burning include significantly increased pH, which Macadam (1987) in the Sub-boreal Spruce Zone of central British Columbia found persisting more than a year after the burn. Average fuel consumption was 20 to 24 t/ha and the forest floor depth was reduced by 28% to 36%. The increases correlated well with the amounts of slash (both total and ≥7 cm diameter) consumed. The change in pH depends on the severity of the burn and the amount consumed; the increase can be as much as 2 units, a 100-fold change. Deficiencies of copper and iron in the foliage of white spruce on burned clearcuts in central British Columbia might be attributable to elevated pH levels.

Even a broadcast slash fire in a clearcut does not give a uniform burn over the whole area. Tarrant (1954), for instance, found only 4% of a 140-ha slash burn had burned severely, 47% had burned lightly, and 49% was unburned. Burning after windrowing obviously accentuates the subsequent heterogeneity.

Marked increases in exchangeable calcium also correlated with the amount of slash at least 7 cm in diameter consumed. Phosphorus availability also increased, both in the forest floor and in the 0 cm to 15 cm mineral soil layer, and the increase was still evident, albeit somewhat diminished, 21 months after burning. However, in another study in the same Sub-boreal Spruce Zone found that although it increased immediately after the burn, phosphorus availability had dropped to below pre-burn levels within 9 months.

Nitrogen will be lost from the site by burning, though concentrations in remaining forest floor were found by Macadam (1987) to have increased in 2 of 6 plots, the others showing decreases. Nutrient losses may be outweighed, at least in the short term, by improved soil microclimate through the reduced thickness of forest floor where low soil temperatures are a limiting factor.

The *Picea/Abies* forests of the Alberta foothills are often characterized by deep accumulations of organic matter on the soil surface and cold soil temperatures, both of which make reforestation difficult and result in a general deterioration in site productivity; Endean and Johnstone (1974) describe experiments to test prescribed burning as a means of seedbed preparation and site amelioration on representative clear-felled *Picea/Abies* areas. Results showed that, in general, prescribed burning did not reduce organic layers satisfactorily, nor did it increase soil temperature, on the sites tested. Increases in seedling establishment, survival, and growth on the burned sites were probably the result of slight reductions in the depth of the organic layer, minor increases in soil temperature, and marked improvements in the efficiency of the planting crews. Results also suggested that the process of site deterioration has not been reversed by the burning treatments applied.

Ameliorative Intervention

Slash weight (the oven-dry weight of the entire crown and that portion of the stem < 4 inches in diameter) and size distribution are major factors influencing the forest fire hazard on harvested sites. Forest managers interested in the application of prescribed burning for hazard reduction and silviculture, were shown a method for quantifying the slash load by Kiil (1968). In west-central Alberta, he felled, measured, and weighed 60 white spruce, graphed (a) slash weight per merchantable unit volume against diameter at breast height (dbh), and (b) weight of fine slash (<1.27 cm) also against dbh, and produced a table of slash weight and size distribution on one acre of a hypothetical stand of white spruce. When the diameter distribution of a stand is unknown, an estimate of slash weight and size distribution can be obtained from average stand diameter, number of trees per unit area, and merchantable cubic foot volume. The sample trees in Kiil's study had full symmetrical crowns. Densely growing trees with short and often irregular crowns would probably be overestimated; open-grown trees with long crowns would probably be underestimated.

The need to provide shade for young outplants of Engelmann spruce in the high Rocky Mountains is emphasized by the U.S. Forest Service. Acceptable planting spots are defined as microsites on the north and east sides of down logs, stumps, or slash, and lying in the shadow cast by such material. Where the objectives of management specify more uniform spacing, or higher densities, than obtainable from an existing distribution of shade-providing material, redistribution or importing of such material has been undertaken.

Access

Site preparation on some sites might be done simply to facilitate access by planters, or to improve access and increase the number or distribution of microsites suitable for planting or seeding.

Wang et al. (2000) determined field performance of white and black spruces 8 and 9 years after outplanting on boreal mixedwood sites following site preparation (Donaren disc trenching versus no trenching) in 2 plantation types (open versus sheltered) in southeastern Manitoba. Donaren trenching slightly reduced the mortality of black spruce but significantly increased the mortality of white spruce. Significant difference in height was found between open and sheltered plantations for black spruce but not for white spruce, and root collar diameter in sheltered plantations was significantly larger than in open plantations for black spruce but not for white spruce. Black spruce open plantation had significantly smaller volume (97 cm^3) compared with black spruce sheltered (210 cm^3), as well as white spruce open (175 cm^3) and sheltered (229 cm^3) plantations. White spruce open plantations also had smaller volume than white spruce sheltered plantations. For transplant stock, strip plantations had a significantly higher volume (329 cm^3) than open plantations (204 cm³). Wang et al. (2000) recommended that sheltered plantation site preparation should be used.

Mechanical

Up to 1970, no "sophisticated" site preparation equipment had become operational in Ontario, but the need for more efficacious and versatile equipment was increasingly recognized. By this time, improvements were being made to equipment originally developed by field staff, and field testing of equipment from other sources was increasing.

According to J. Hall (1970), in Ontario at least, the most widely used site preparation technique was post-harvest mechanical scarification by equipment front-mounted on a bulldozer (blade, rake, V-plow, or teeth), or dragged behind a tractor (Imsett or S.F.I. scarifier, or rolling chopper). Drag type units designed and constructed by Ontario's Department of Lands and Forests used anchor chain or tractor pads separately or in combination, or were finned steel drums or barrels of various sizes and used in sets alone or combined with tractor pad or anchor chain units.

J. Hall's (1970) report on the state of site preparation in Ontario noted that blades and rakes were found to be well suited to post-cut scarification in tolerant hardwood stands for natural regeneration of yellow birch. Plows were most effective for treating dense brush prior to planting, often in conjunction with a planting machine. Scarifying teeth, e.g., Young's teeth, were sometimes used to prepare sites for planting, but their most effective use was found to be preparing sites for seeding, particularly in backlog areas carrying light brush and dense herbaceous growth. Rolling choppers found application in treating heavy brush but could be used only on stone-free soils. Finned drums were commonly used on jack pine–spruce cutovers on fresh brushy sites with a deep duff layer and heavy slash, and they needed to be teamed with a tractor pad unit to secure good distribution of the slash. The S.F.I. scarifier, after strengthening, had been "quite successful" for 2 years, promising trials were under way with the cone scarifier and barrel ring scarifier, and development had begun on a new flail scarifier for use on sites with shallow, rocky soils. Recognition of the need to become more effective and efficient in site preparation led the Ontario Department of Lands and Forests to adopt the policy of seeking and obtaining for field testing new equipment from Scandinavia and elsewhere that seemed to hold promise for Ontario conditions, primarily in the north. Thus, testing was begun of the Brackekultivator from Sweden and the Vako-Visko rotary furrower from Finland.

Mounding

Site preparation treatments that create raised planting spots have commonly improved outplant performance on sites subject to low soil temperature and excess soil moisture. Mounding can certainly have a big influence on soil temperature. Draper et al. (1985), for instance, documented this as well as the effect it had on root growth of outplants (Table 30).

The mounds warmed up quickest, and at soil depths of 0.5 cm and 10 cm averaged 10 and 7 °C higher, respectively, than in the control. On sunny days, daytime surface temperature maxima on the mound and organic mat reached 25 °C to 60 °C, depending on soil wetness and shading. Mounds reached mean soil temperatures of 10 °C at 10 cm depth 5 days after planting, but the control did not reach that temperature until 58 days after planting. During the first growing season, mounds had 3 times as many days with a mean soil temperature greater than 10 °C than did the control microsites.

Draper et al.'s (1985) mounds received 5 times the amount of photosynthetically active radiation (PAR) summed over all sampled microsites throughout the first growing season; the control treatment consistently received about 14% of daily background PAR, while mounds received over 70%. By November, fall frosts had reduced shading, eliminating the differential. Quite apart from its effect on temperature, incident radiation is also important photosynthetically. The average control microsite was exposed to levels of light above the compensation point for only 3 hours, i.e., one-quarter of the daily light period, whereas mounds received light above the compensation point for 11 hours, i.e., 86% of the same daily period. Assuming that incident light in the 100-600 $\mu Em^{-2}s^{-}1$ intensity range is the most important for photosynthesis, the mounds received over 4 times the total daily light energy that reached the control microsites.

Orientation of Linear Site Preparation, e.g., Disk-Trenching

With linear site preparation, orientation is sometimes dictated by topography or other considerations, but the orientation can often be chosen. It can make a difference. A disk-trenching experiment in the Sub-boreal Spruce Zone in interior British Columbia investigated the effect on growth of young outplants (lodgepole pine) in 13 microsite planting positions: berm, hinge, and trench in each of north, south, east, and west aspects, as well as in untreated locations between the furrows. Tenth-year stem volumes of trees on south, east, and west-facing microsites were significantly greater than those of trees on north-facing and untreated microsites. However, planting spot selection was seen to be more important overall than trench orientation.

In a Minnesota study, the N–S strips accumulated more snow but snow melted faster than on E–W strips in the first year after felling. Snow-melt was faster on strips near the centre of the strip-felled area than on border strips adjoining the intact stand. The strips, 50 feet (15.24 m) wide, alternating with uncut strips 16 feet (4.88 m) wide, were felled in a *Pinus resinosa* stand, aged 90 to 100 years.

SOIL CONSERVATION

Soil conservation is the prevention of soil loss from erosion or reduced fertility caused by over usage, acidification, salinization or other chemical soil contamination.

Slash-and-burn and other unsustainable methods of subsistence farming are practiced in some lesser developed areas. A sequel to the deforestation is typically large scale erosion, loss of soil nutrients and sometimes total desertification.

Erosion barriers on disturbed slope, Marin County, California

Techniques for improved soil conservation include crop rotation, cover crops, conservation tillage and planted windbreaks and affect both erosion and fertility. When plants, especially trees, die, they decay and become part of the soil. Code 330 defines standard methods recommended by the US Natural Resources Conservation Service.

Contour plowing, Pennsylvania 1938. The rows formed slow water run-off during rainstorms to prevent soil erosion and allows the water time to infiltrate into the soil.

Farmers have practiced soil conservation for millennia. Conservation practices fall in multiple categories:

Contour Plowing

Contour plowing orients crop furrows following the contour lines of the farmed area. Furrows move left and right to maintain a constant altitude, which reduces runoff. Contour plowing was practiced by the ancient Phoenicians, and is effective for slopes between two and ten percent. Contour plowing can increase crop yields from 10 to 50 percent, partially as a result from greater soil retention.

Terracing or terrace farming

Terracing is the practice of creating nearly level areas in a hillside area. The terraces form a series of steps, each at a higher level than the previous. Terraces are protected from erosion by other soil barriers.

Terraced farming is more common on small farms and in underdeveloped countries, since mechanized equipment is difficult to deploy in this setting. It protects the soil from its erosion. It is one of the way by which soil erosion can be stopped.

Keyline Design

Keyline design is an enhancement of contour farming, where the total watershed properties are taken into account in forming the contour lines.

Perimeter Runoff Control

runoff and filter soxx

Tree, shrubs and ground-cover are effective perimeter treatment for soil erosion prevention, by impeding surface flows. A special form of this perimeter or inter-row treatment is the use of a “grass way” that both channels and dissipates runoff through surface friction, impeding surface runoff and encouraging infiltration of the slowed surface water.

Windbreaks

Windbreaks are sufficiently dense rows of trees at the windward exposure of an agricultural field subject to wind erosion. Evergreen species provide year-round protection; however, as long as foliage is present in the seasons of bare soil surfaces, the effect of deciduous trees may be adequate.

Cover Crops/Crop Rotation

Cover crops such as legumes, white turnip, radishes and other species are rotated with cash crops to blanket the soil year-round and act as green manure that replenishes nitrogen and other critical nutrients. Cover crops also help suppress weeds.

Soil-Conservation Farming

Soil-conservation farming involves no-till farming, "green manures" and other soil-enhancing practices. Such farming methods attempt to mimic the biology of virgin land. They can revive damaged soil, minimize erosion, encourage plant growth, eliminate the use of nitrogen fertilizer or fungicide, produce above-average yields and protect crops during droughts or flooding. The result is less labor and lower costs that increase farmers' profits. No-till farming and cover crops act as sinks for nitrogen and other nutrients. This increases the amount of soil organic matter.

Repeated plowing/tilling degrades soil, killing its beneficial fungi and earthworms. Once damaged, soil may take multiple seasons to fully recover, even in optimal circumstances.

Critics argue that no-till and related methods are impractical and too expensive for many growers, partly because it requires new equipment. They cite advantages for conventional tilling depending on the geography, crops and soil conditions. Some farmers claimed that no-till complicates weed control, delays planting and that post-harvest residues, especially for corn, are hard to manage.

Salinity Management

Salt deposits on the former bed of the Aral Sea

Salinity in soil is caused by irrigating with salty water. Water then evaporates from the soil leaving the salt behind. Salt breaks down the soil structure, causing infertility and reduced growth.

The ions responsible for salination are: sodium (Na^{+}), potassium (K^{+}), calcium (Ca^{2+}), magnesium (Mg^{2+}) and chlorine (Cl^{-}). Salinity is estimated to affect about one third of the earth's arable land. Soil salinity adversely affects crop metabolism and erosion usually follows.

Salinity occurs on drylands from overirrigation and in areas with shallow saline water tables. Over-irrigation deposits salts in upper soil layers as a byproduct of soil infiltration; irrigation merely increases the rate of salt deposition. The best-known case of shallow saline water table capillary action occurred in Egypt after the 1970 construction of the Aswan Dam. The change in the groundwater level led to high salt concentrations in the water table. The continuous high level of the water table led to soil salination.

Use of humic acids may prevent excess salination, especially given excessive irrigation. Humic acids can fix both anions and cations and eliminate them from root zones. Planting species that can tolerate saline conditions to produce surface cover can preserve soil salinity can be reduced. Salt-tolerant plants include saltbush, a plant found in much of North America and in the Mediterranean regions of Europe.

Soil Organisms

Yellow fungus, a mushroom that assists in organic decay.

When worms excrete egesta in the form of casts, a balanced selection of minerals and plant nutrients is made into a form accessible for root uptake. Earthworm casts are five times richer in available nitrogen, seven times richer in available phosphates and eleven times richer in available potash than the surrounding upper 150 millimetres (5.9 in) of soil. The weight of casts produced may be greater than 4.5 kg per worm per year. By burrowing, the earthworm improves soil porosity, creating channels that enhance the processes of aeration and drainage.

Other important soil organisms include nematodes, mycorrizha and bacteria.

Degraded soil requires synthetic fertilizer to produce high yields. Lacking structure increases erosion and carries nitrogen and other pollutants into rivers and streams.

Each one percent increase in soil organic matter helps soil hold **20,000** gallons more water per acre.

Mineralization

To allow plants full realization of their phytonutrient potential, active mineralization of the soil is sometimes undertaken. This can involve adding crushed rock or chemical soil supplements. In either case the purpose is to combat mineral depletion. A broad range of minerals can be used, including common substances such as phosphorus and more exotic substances such as zinc and selenium. Extensive research examines the phase transitions of minerals in soil with aqueous contact.

Flooding can bring significant sediment to an alluvial plain. While this effect may not be desirable if floods endanger life or if the sediment originates from productive land, this process of addition to a floodplain is a natural process that can rejuvenate soil chemistry through mineralization.

Soil Management

Soil management concerns all operations, practices, and treatments used to protect soil and enhance its performance (such as soil fertility or soil mechanics). It includes soil conservation, soil amendment, and optimal soil health. In agriculture, some amount of soil management is needed both in nonorganic and organic types to prevent agricultural land from becoming poorly productive over decades. Organic farming in particular stresses optimal soil management, because it uses soil health as the exclusive or nearly exclusive source of its fertilization and pest control.

Environment

According to the EPA, agricultural soil management practices can lead to production and emission of nitrous oxide (N_2O), a major greenhouse gas and air pollutant. Activities that can contribute to N2O emissions include fertilizer usage, irrigation and tillage. The management of soils accounts for over half of the emissions from the Agriculture sector. Cattle livestocks account for one third of emissions, through methane emissions. Manure management and rice cultivation also emit emissions.

Methods that significantly enhance carbon sequestration in soil include no-till farming, residue mulching, cover cropping, and crop rotation, all of which are more widely used in organic farming than in conventional farming. Because only 5% of US farmland currently uses no-till and residue mulching, there is a large potential for carbon sequestration.

Practices

Soil management practices that affect soil quality:

- Controlling traffic on the soil surface helps to reduce soil compaction, which can reduce aeration and water infiltration.
- Cover crops keep the soil anchored and covered in off-seasons so that the soil is not eroded by wind and rain.
- Crop rotations for row crops alternate high-residue crops with lower-residue crops to increase the amount of plant material left on the surface of the soil during the year to protect the soil from erosion.
- Nutrient management can help to improve the fertility of the soil and the amount of organic matter content, which improves soil structure and function.
- Tillage, especially reduced-tillage or no-till operations limit the amount of soil disturbance while cultivating a new crop, and help to maintain plant residues on the surface of the soil for erosion protection and water retention.

Advantages of Soil Management

- Maintain soil fertility
- Restore soil fertilitys
- Make the agricultural process an economic one
- Help increase yield

No-Till Farming

Young soybean plants thrive in and are protected by the residue of a wheat crop. This form of no till farming provides good protection for the soil from erosion and helps retain moisture for the new crop.

No-till farming (also called **zero tillage** or **direct drilling**) is a way of growing crops or pasture from year to year without disturbing the soil through tillage. No-till is an agricultural technique which increases the amount of water that infiltrates into the soil and increases organic matter retention and cycling of nutrients in the soil. In many agricultural regions it can reduce or eliminate soil erosion. It increases the amount and variety of life in and on the soil, including disease-causing organisms and disease suppression organisms. The most powerful benefit of no-tillage is improvement in soil biological fertility, making soils more resilient. Farm operations are made much more efficient, particularly improved time of sowing and better trafficability of farm operations.

Tillage remains relevant in agriculture today, but the success of no-till methods in many contexts keeps farmers aware that multiple options exist. In some cases **low-till** methods combine aspects of till and no-till methods. For example, some approaches may use a limited amount of shallow disc harrowing but no plowing.

Tilling is the process of removing plants or plant debris, usually for the purposes of planting more desirable species. This tilling can result in a flat seed bed or one that has formed areas, such as rows or raised beds, to enhance the growth of desired plants. It is an ancient technique with clear evidence of its use since at least 3000 B.C.

The effects of tillage can include soil compaction; loss of organic matter; degradation of soil aggregates; death or disruption of soil microbes and other organisms including mycorrhizae, arthropods, and earthworms; and soil erosion where topsoil is washed or blown away.

Origin of No-till for Modern Farms

The idea of modern no-till farming started in the 1940s with Edward H. Faulkner, author of *Plowman's Folly*, but it wasn't until the development of several chemicals after WWII that various researchers and farmers started to try out the idea. The first adopters of no-till include Klingman (North Carolina), Edward Faulkner, L.A. Porter (New Zealand), Harry and Lawrence Young (Herndon, Kentucky), the Instituto de Pesquisas Agropecuarias Meridional (1971 in Brazil) with Herbert Bartz.

Adoption rate in the United States

No-till farming is widely used in the United States and the number of acres managed in this way continues to grow. This growth is supported by a decrease in costs related to tillage; no-till management results in fewer passes with equipment for approximately equal harvests, and the crop residue prevents evaporation of rainfall and increases water infiltration into the soil.

Issues

Profit, Economics, Yield

Studies have found that no-till farming can be more profitable if performed correctly.

Less tillage of the soil reduces labour, fuel, irrigation and machinery costs. No-till can increase yield because of higher water infiltration and storage capacity, and less erosion. Another benefit of no-till is that because of the higher water content, instead of leaving a field fallow it can make economic sense to plant another crop instead.

As sustainable agriculture becomes more popular, monetary grants and awards are becoming readily available to farmers who practice conservation tillage. Some large energy corporations which are among the greatest generators of fossil-fuel-related pollution may purchase carbon credits, which can encourage farmers to engage in conservation tillage. Under such schemes, the farmers` land is legally redefined as a carbon sink for the power generators` emissions. This helps the farmer in several ways, and it helps the energy companies meet regulatory demands for reduction of pollution, specifically carbon emissions.

No-till farming can increase organic (carbon based) matter in the soil, which is a form of carbon sequestration. However, there is debate over whether this increased sequestration detected in scientific studies of no-till agriculture is actually occurring, or is due to flawed testing methods or other factors. Regardless of this debate, a case can still be made for no-till, in the form of reduction in fossil fuel use, less erosion and better soil quality.

Environmental

Carbon (air and soil) and other Greenhouse Gases

No-till farming has carbon sequestration potential through storage of soil organic matter in the soil of crop fields. Whereas, when soil is tilled by machinery, the soil layers invert, air mixes in, and soil microbial activity dramatically increases over baseline levels. Tilling results in soil organic matter being broken down much more rapidly, and carbon is lost from the soil into the atmosphere. In addition to the effect on soil from tilling, emissions from the farm tractors increases carbon dioxide levels in the atmosphere.

Cropland soils are ideal for use as a carbon sink, since they have been depleted of carbon in most areas. It is estimated that 78 billion metric tonnes of carbon that was trapped in the soil has been released because of tillage. Conventional farming practices that rely on tillage have removed carbon from the soil ecosystem by removing crop residues such as left over corn stalks, and through the addition of chemical fertilizers which have the above-mentioned effects on soil microbes. By eliminating tillage, crop residues decompose where they lie, and by growing winter cover crops, carbon loss can be slowed and eventually reversed.

Nonetheless, a growing body of research is showing that no-till systems lose carbon stocks over time. Regarding a 2014 study of which he was principal investigator, University of Illinois soil scientist Ken Olson said this differing result occurs in part because tested soil samples need to include the full depth of rooting; 1–2 meters deep. He said, "That no-till subsurface layer is often losing more soil organic carbon stock over time than is gained in the surface layer." Also, there has not been a uniform definition of soil organic carbon sequestration among researchers. The study concludes, Additional investments in SOC research is needed to better understand the agricultural management practices that are most likely to sequester SOC or at least retain more net SOC stocks."

In addition to keeping carbon in the soil, no-till farming reduces nitrous oxide (N_2O) emissions by 40-70%, depending on rotation. Nitrous oxide is a potent greenhouse gas, 300 times stronger than CO2, and stays in the atmosphere for 120 years. Fertilizing farmlands with (excessive) nitrogen increases the release of nitrous oxide.

Soil and Water

No-till farming improves soil quality (soil function), carbon, organic matter, aggregates, protecting the soil from erosion,evaporation of water, and structural breakdown. A reduction in tillage passes helps prevent the compaction of soil.

Recently, researchers at the Agricultural Research Service of the United States Department of Agriculture found that no-till farming makes soil much more stable than plowed soil. Their conclusions draw from over 19 years of collaborated tillage studies. No-till stores more carbon in the soil and carbon in the form of organic matter is a key factor in holding soil particles together. The first inch of no-till soil is two to seven times less vulnerable than that of plowed soil. The practice of no-till farming is especially beneficial to Great Plains farmers because of its avoidance of erosion.

Crop residues left intact help both natural precipitation and irrigation water infiltrate the soil where it can be used. The crop residue left on the soil surface also limits evaporation, conserving water for plant growth. Soil compaction and no tillage-pan, soil absorbs more water and plants are able to grow their roots deeper into the soil and suck up more water.

Tilling a field reduces the amount of water, via evaporation, around 1/3 to 3/4 inches (0.85 to 1.9 cm) per pass. By no-tilling, this water stays in the soil, available to the plants.

Soil Biota, Wildlife, etc.

In no-till farming the soil is left intact and crop residue is left on the field. Therefore, soil layers, and in turn soil biota, are conserved in their natural state. No-tilled fields often have more beneficial insects and annelids, a higher microbial content, and a greater amount of soil organic material. Since there is no ploughing there is less airborne dust.

No-till farming increases the amount and variety of wildlife. This is the result of improved cover, reduced traffic and the reduced chance of destroying ground nesting birds and animals (plowing destroys all of them).

Albedo

Tillage lowers the albedo of croplands. The potential for global cooling as a result of increased Albedo in no till croplands is similar in magnitude to the biogeochemical (carbon sequestration) potential.

Historical Artifacts

Tilling regularly damages ancient structures under the soil such as long barrows. In the UK, half of the long barrows in Gloucestershire and almost all the burial mounds in

Essex have been damaged. According to English Heritage modern tillage techniques have done as much damage in the last six decades as traditional tilling did in the previous six centuries. By using no-till methods these structures can be preserved and can be properly investigated instead of being destroyed.

Cost

Equipment

No-till farming requires specialized seeding equipment designed to plant seeds into undisturbed crop residues and soil. If the farmer has equipment designed for tillage farming, purchasing new equipment (seed drills for example) would be expensive and while the cost could be offset by selling off plows, etc. doing so is not usually done until the farmer decides to switch completely over (after trying it out for a few years). This would result in more money being invested into equipment in the short term (until old equipment is sold off).

Drainage

If a soil has poor drainage, it may need drainage tiles or other devices in order to help with the removal of excess water under no-till. Farmers should remember that water infiltration will improve after several years of a field being in no-till farming, so they may want to wait until 5–8 years have passed to see if the problems persists before deciding to invest in such an expensive system.

Gullies

Gullies can be a problem in the long-term. While much less soil is displaced by using no-till farming, any drainage gulleys that do form will get deeper each year since they aren't being smoothed out by plowing. This may necessitate either sod drainways, waterways, permanent drainways, cover crops, etc. Gully formation can be avoided entirely with proper water management practices, including the creation of swales on contour.

Increased Chemical Use

One of the purposes of tilling is to remove weeds. No-till farming does change weed composition drastically. Faster growing weeds may no longer be a problem in the face of increased competition, but shrubs and trees may begin to grow eventually.

Some farmers attack this problem with a "burn-down" herbicide such as glyphosate in lieu of tillage for seedbed preparation and because of this, no-till is often associated with increased chemical use in comparison to traditional tillage based methods of crop production. However, there are many agroecological alternatives to increased chemical use, such as winter cover crops and the mulch cover they provide, soil solarization or burning.

Management

No-till farming requires some different skills to achieve success. As with any production system, if no-till isn't done correctly, yields can drop. A combination of technique, equipment, pesticides, crop rotation, fertilization, and irrigation have to be used for local conditions.

Cover Crops

Cover crops are used occasionally in no-till farming to help control weeds and increase nutrients in the soil (by using legumes) or by using plants with long roots to pull mobile nutrients back up to the surface from lower layers of the soil. Farmers experimenting with organic no-till use cover crops instead of tillage for controlling weeds, and are developing various methods to kill the cover crops (rollers, crimper, choppers, etc.) so that the newly planted crops can get enough light, water, nutrients, etc.

Disease, Pathogens, Insects and the use of Crop Rotations

With no-till farming, residue from the previous years crops lie on the surface of the field, cooling it and increasing the moisture. This can cause increased or decreased or variations of diseases that occur, but not necessarily at a higher or lower rate than conventional tillage. In order to help eliminate weed, pest and disease problems, crop rotations are used. By rotating the crops on a multi-year cycle, pests and diseases will decrease since the pests will no longer have a food supply to support their numbers.

Organic no-till Technique: The Cardboard Method

Some farmers who prefer to pursue a chemical-free management practice often rely on the use of normal, non-dyed corrugated cardboard for use on seed-beds and vegetable areas. Used correctly, cardboard placed on a specific area can

- keep important fungal hyphae and microorganisms in the soil intact
- prevent recurring weeds from popping up
- increase residual nitrogen and plant nutrients by top-composting plant residues and
- create valuable topsoil that is well suited for next years seeds or transplants.

The plant residues (left over plant matter originating from cover crops, grass clippings, original plant life etc.) will rot while underneath the cardboard so long as it remains sufficiently moist. This rotting attracts worms and other beneficial microorganisms to the site of decomposition, and over a series of a few seasons (usually Spring-->Fall or Fall->Spring) and up to a few years, will create a layer of rich topsoil. Plants can then be direct seeded into the soil come spring, or holes can be cut into the cardboard to allow for transplantation. Using this method in conjunction with other sustainable practices such as composting/vermicompost, cover crops and rotations are often considered beneficial to both land and those who take from it.

Water Issues

No-till farming dramatically reduces the amount of erosion in a field. While much less soil is displaced, any gullies that do form will get deeper each year instead of being smoothed out by regular plowing. This may necessitate either sod drainways, waterways, permanent drainways, cover crops, etc.

A problem that occurs in some fields is water saturation in soils. Switching to no-till farming will correct the drainage the because of the qualities of soil under continuous no-till include a higher water infiltration rate.

Equipment

It is very important to have planting equipment that can properly penetrate through the residue, into the soil and prepare a good seedbed. Switching to no-till reduces the maximum amount of power needed from farm tractors, which means that a farmer can farm under no-till with a smaller tractor than under tilling. Using a smaller, lighter tractor has the added benefit of reducing compaction.

Soil Temperature

Another problem that growers face is that in the spring the soil will take longer to warm and dry, which may delay planting to a less ideal future date. One reason why the soil is slower to warm is that the field absorbs less solar energy as the residue covering the soil is a much lighter color than the black soil which would be exposed in conventional tillage. This can be managed by using row cleaners on a planter. Since the soil can be cooler, harvest can occur a few days later than a conventionally tilled field. Note: A cooler soil is also a benefit because water doesn't evaporate as fast.

Residue

On some crops, like continuous no-till corn, the thickness of the residue on the surface of the field can become a problem without proper preparation and/or equipment.

Fertilizer

One of the most common yield reducers is nitrogen being immobilized in the crop residue, which can take a few months to several years to decompose, depending on the crop's C to N ratio and the local environment. Fertilizer needs to be applied at a higher rate during the transition period while the soil rebuilds its organic matter. The nutrients in the organic matter will be eventually released back into the soil, so this is only a concern during the transition time frame (4–5 years for Kansas, USA). An innovative solution to this problem is to integrate animal husbandry in various ways to aid in the decomposition cycle.

Misconceptions

Need to fluff the soil

Although no-till farming often causes a slight increase in soil bulk density, periodic tilling is not needed to "fluff" the soil back up. No-till farming mimics the natural conditions under which most soils formed more closely than any other method of farming, in that the soil is left undisturbed except to place seeds in a position to germinate.

Similar Terms

No-till farming is not equivalent to conservation tillage or strip tillage. Conservation tillage is a group of practices that reduce the amount of tillage needed. No-till and strip tillage are both forms of conservation tillage. No-till is the practice of never tilling a field. Tilling every other year is called rotational tillage.

REFERENCES

1. Predicting soil erosion by water, a guide to conservation planning in the Revised Universal Soil Loss Equation, United States Department of Agriculture, Agricultural Research Service, Agricultural handbook no. 703 (1997)
2. Contour farming boosts yields : a farmer's guide in laying out key contour lines and establishing grassed waterways. United States. 1943. p. 1. Retrieved 1 August 2015. |first1= missing |last1= in Authors list (help)
3. Perimeter landscaping of Carneros Business Park, Lumina Technologies, Santa Rosa, Ca., prepared for Sonoma County, Ca. (2002)
4. Wolfgang Summer, Modelling Soil Erosion, Sediment Transport and Closely Related Hydrological Processes entry by Mingyuan Du, Peiming Du, Taichi Maki and Shigeto Kawashima, "Numerical modeling of air flow over complex terrain concerning wind erosion", International Association of Hydrological Sciences publication no. 249 (1998) ISBN 1-901502-50-3
5. Goode, Erica (March 10, 2015). Farmers Put Down the Plow for More Productive Soil". The New York Times (New York ed.). The New York Times Company. p. D1. ISSN 0362-4331. OCLC 1645522. Retrieved April 5, 2015.
6. Dan Yaron, Salinity in Irrigation and Water Resources, Marcel Dekker, New York (1981) ISBN 0-8247-6741-1
7. Bill Mollison, Permaculture: A Designer`s Manual, Tagari Press, (December 1, 1988), 576 pages, ISBN 0908228015. Increases in porosity enhance infiltration and thus reduce adverse effects of surface runoff.
8. Arthur T. Hubbard, Encyclopedia of Surface and Colloid Science Vol 3, Santa Barbara, California Science Project, Marcel Dekker, New York (2004) ISBN 0-8247-0759-1

9. Agriculture: Sources of Greenhouse Gas Emissions". EPA. 2015.
10. Susan S. Lang (13 July 2005). Organic farming produces same corn and soybean yields as conventional farms, but consumes less energy and no pesticides, study finds". Retrieved 8 July 2008.
11. Pimentel, David; Hepperly, Paul; Hanson, James; Douds, David; Seidel, Rita (2005). Environmental, Energetic, and Economic Comparisons of Organic and Conventional Farming Systems". BioScience. 55 (7): 573–82. doi0.1641/0006-3568(2005)055[0573:EEAECO]2.0.CO;2.
12. Lal, Rattan; Griffin, Michael; Apt, Jay; Lave, Lester; Morgan, M. Granger (2004). Ecology: Managing Soil Carbon". Science. 304 (5669): 393. doi0.1126/science.1093079. PMID 15087532.
13. Soilquality.org - Soil Management Practices
14. Soil Management - Penn State

CHAPTER 7

Livestock Management Practices

Chapter Contents

The word "livestock" is an umbrella term used for domesticated animals raised in an agricultural environment, with the intent of providing food, textiles, labor, or fertilizer to their owners. Common examples are horses, pigs, goats, cows, sheep, and poultry, although numerous other semi-wild animals including reindeer, yaks, camels, and emus could also be considered livestock. Humans have coexisted with domesticated animals for centuries, and the rise of farming and keeping animals probably contributed to a major shift in human culture.

INTRODUCTION

Livestock are domesticated animals raised in an agricultural setting to produce commodities such as food, fiber, and labor. The term is often used to refer solely to those raised for food, and sometimes only farmed ruminants, such as cattle and goats. In recent years, some organizations have also raised livestock to promote the survival of rare breeds. The breeding, maintenance, and slaughter of these animals, known as animal husbandry, is a component of modern agriculture that has been practiced in many cultures since humanity's transition to farming from hunter-gatherer lifestyles.

Cattle on a pasture in Germany.

Animal husbandry practices have varied widely across cultures and time periods. Originally, livestock were not confined by fences or enclosures, but these practices have largely shifted to intensive animal farming, sometimes referred to asfactory farming. These practices increase yield of the various commercial outputs, but have led to increased concerns about animal welfare and environmental impact. Livestock production continues to play a major economic and cultural role in numerous rural communities.

ETYMOLOGY AND LEGAL DEFINITION

This Australian road sign uses the less common term "stock" for livestock.

Livestock as a word was first used between 1650 and 1660, as a merger between the words "live" and "stock".

Older English sources, such as the King James Version of the Bible, refer to all domesticated animals as "cattle", while the word "deer" was used for wild animals. The word cattle is derived from Old North French *catel*, which meant all kinds of movable personal property, including livestock, which was differentiated from immovable real estate ("real property"). In later English, sometimes smaller livestock such as chickens and pigs were referred to as "small cattle".. Today, the modern meaning of cattle, without a modifier, usually refers to domesticated bovines, but sometimes livestock refers only to this subgroup.

Legal Definition

United States federal legislation sometimes more narrowly defines the term to make specified agricultural commodities either eligible or ineligible for a program or activity. For example, the Livestock Mandatory Reporting Act of 1999 (P.L. 106-78, Title IX) defines livestock only as cattle, swine, and sheep. The 1988 disaster assistance legislation defined the term as "cattle, sheep, goats, swine, poultry (including egg-producing poultry), equine animals used for food or in the production of food, fish used for food, and other animals designated by the Secretary."

HISTORY

Animal-rearing originated during the cultural transition to settled farming communities from hunter-gatherer lifestyles. Animals are domesticated when their breeding and living conditions are controlled by humans. Over time, the collective behaviour, lifecycle and physiology of livestock have changed radically. Many modern farm animals are unsuited to life in the wild.

Dogs were domesticated in East Asia about 15,000 years ago. Goats and sheep were domesticated around 8000 BC in Asia. Swine or pigs were domesticated by 7000 BC in the Middle East and China. The earliest evidence of horse domestication dates to around 4000 BC.

TYPES

The term "livestock" is nebulous and may be defined narrowly or broadly. Broadly, livestock refers to any breed or population of animal kept by humans for a useful, commercial purpose. This can mean domestic animals, semidomestic animals, or captive wild animals. Semidomesticated refers to animals which are only lightly domesticated or of disputed status. These populations may also be in the process of domestication. Some people may use the term livestock to refer to only animals used for red meat.

Animal / Type	Domestication status	Wild ancestor	Time of first captivity, domestication	Area of first captivity, domestication	Current commercial uses	Picture	Ref
Alpaca Mammal, herbivore	domestic	Vicuña	Between 5000 BC and 4000 BC	Andes	Alpaca fiber, meat		
Addax Mammal, herbivore	domestic	Addax	2500 BCE	Egypt	Meat, hides		
Bali cattle Mammal, herbivore	domestic	Banteng	Unknown	Southeast Asia, Bali	Meat, milk, draught		
Bison Mammal, herbivore	captive (see also Beefalo)	N/A	Late 19th century	North America	Meat, leather		
Camel Mammal, herbivore	domestic	Wild dromedary and Bactrian camels	Between 4000 BC and 1400 BC	Asia	Mount, pack animal, meat, milk, camel hair		
Cattle Mammal, herbivore	domestic	Aurochs	6000 BC	Southwest Asia, South Asia, North Africa	Meat (beef, veal, blood), milk, leather, draught		
Capybara Mammal, herbivore	captive	Capybara	Unknown	South America	Meat, skins, pets		
Collared peccary Mammal, omnivore	captive	Collared peccary	Unknown	Brazil	Meat, tusks, skins, pets		

Deer Mammal, herbivore	captive	N/A	First century AD	UK	Meat (venison), leather, antlers, antler velvet		
Donkey Mammal, herbivore	domestic	African wild ass	4000 BC	Egypt	Mount, pack animal, draught, meat, milk		
Eland Mammal, herbivore	domestic	Common eland, Giant eland	Unknown	South Africa, Kenya, Zimbabwe, West Africa	Meat, milk, leather, hides, horns		
Elk Mammal, herbivore	captive	Elk	1990s	North America	Meat, antlers, leather, hides		
Fallow deer Mammal, herbivore	semidomestic	Fallow deer	9th century BC	Mediterranean Basin	Meat, antlers, hides, ornamentation		
Gayal Mammal, herbivore	domestic	Gaur	Unknown	Southeast Asia	Meat, draught		
Goat Mammal, herbivore	domestic	Wild goat	8000 BC	Southwest Asia	Milk, meat, wool, leather, light draught		
Guinea pig Mammal, herbivore	domestic	*Cavia tschudii*	5000 BC	South America	Meat, pet companionship		
Greater cane rat Mammal, herbivore	captive	Greater cane rat	Unknown	West Africa	Meat	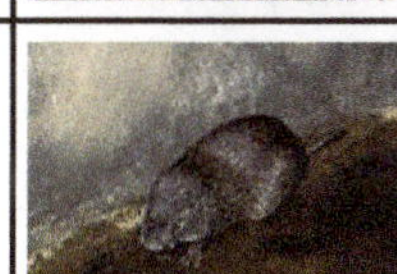	

Greater kudu Mammal, herbivore	captive	Greater kudu	Unknown	South Africa	Meat, hides, horns, leather, pet		
Horse Mammal, herbivore	domestic	Wild horse	4000 BC	Eurasian Steppes	Mount, draught, milk, meat, pack animal		
Llama Mammal, herbivore	domestic	Guanaco	3500 BC	Andes	Pack animal, draught, meat, fiber		
Mule Mammal, herbivore	domestic	Sterile hybrid of donkey and horse			Mount, pack animal, draught		
Moose Mammal, herbivore	domestic	Moose	1940s	Russia, Sweden, Finland, Alaska	Meat, milk, antlers, research, draft		
Muskox Mammal, herbivore	domestic	Muskox	1960s	Alaska	Meat, wool, milk		
Pig Mammal, omnivore	domestic	Wild boar	7000 BC	Eastern Anatolia	Meat (pork), leather, pet, research		
Rabbit Mammal, herbivore	domestic	Wild rabbit	AD 400-900	France	Meat, fur, leather, pet, research		
Reindeer Mammal, herbivore	semidomestic	Reindeer	3000 BC	Northern Russia	Meat, leather, antlers, milk, draught		

Sika deer Mammal, herbivore	domestic	Sika deer	Unknown	Japan, China	Meat, antlers, hides, leather, pet, tourism		
Scimitar oryx Mammal, herbivore	domestic	Scimitar oryx	2320-2150 BC	Egypt	Meat, sacrifice, horns, hides, leather		
Sheep Mammal, herbivore	domestic	Asiatic mouflon sheep	Between 11000 and 9000 BC	Southwest Asia	Wool, milk, leather, meat (lamb and mutton)		
Thorold's deer Mammal, herbivore	captive	Thorold's deer	Unknown	China	Meat, antlers		
White-tailed deer Mammal, herbivore	captive	White-tailed deer	Unknown	West Virginia, Florida, Colombia	Meat, antlers, hides, companionship		
Water buffalo Mammal, herbivore	domestic	Wild Asian water buffalo (Arni)	4000 BC	South Asia	Mount, draught, meat, milk		
Yak Mammal, herbivore	domestic	Wild Yak	2500 BC	Tibet, Nepal	Meat, milk, fiber, mount, pack animal, draught		
Zebu Mammal, herbivore	domestic	Aurochs	8000 BC	India	Meat, milk, draught, hides		

ANIMAL REARING

A Brown Swiss cow in the Swiss Alps.

Livestock are used by humans for a variety of purposes, many of which have an economic value. Livestock products include:

Meat

A useful form of dietary protein and energy, meat is the edible tissue of the animal carcass.

Dairy Products

Mammalian livestock can be used as a source of milk, which can in turn easily be processed into other dairy products, such as yogurt, cheese, butter, ice cream, kefir, and kumis. Using livestock for this purpose can often yield several times the food energy of slaughtering the animal outright.

Clothing and Adornment

Livestock produce a range of fiber textiles. For example, domestic sheep and goats produce wool and mohair, respectively; cattle, swine, deer, and sheep skins can be made into leather; livestock bones, hooves and horns can be used to fabricate jewellery, pendants, or headgear.

Fertilizer

Manure can be spread on fields to increase crop yields. This is an important reason why historically, plant and animal domestication have been intimately linked. Manure is also used to make plaster for walls and floors, and can be used as a fuel for fires. The blood and bone of animals are also used as fertilizer.

Labor

The muscles of animals such as horses, donkeys, and yaks can be used to provide mechanical work. Prior to steam power, livestock were the only available source of nonhuman labor. They are still used in many places of the world to plough fields (drafting), transport goods and people, in military functions, and to power treadmills for grinding grain.

Land Management

The grazing of livestock is sometimes used as a way to control weeds and undergrowth. For example, in areas prone to wildfires, goats and sheep are set to graze on dry scrub which removes combustible material and reduces the risk of fires.

Conservation

The raising of livestock to conserve a rare breed can be achieved through gene banking and breeding programmes.

During the history of animal husbandry, many secondary products have arisen in an attempt to increase carcass utilization and reduce waste. For example, animal offal and inedible parts may be transformed into products such as pet food and fertilizer. In the past, such waste products were sometimes also fed to livestock, as well. However, intraspecies recycling poses a disease risk, threatening animal and even human health (see bovine spongiform encephalopathy (BSE), scrapie, and prion). Due primarily to BSE (mad cow disease), feeding animal scraps to animals has been banned in many countries, at least for ruminants.

FARMING PRACTICES

Goat family with 1-week-old kid

Farrowing site in a natural cave in northern Spain.

Farming practices vary dramatically worldwide and among types of animals. Livestock are generally kept in an enclosure, fed by humans, and intentionally bred. However, some livestock are not enclosed, are fed by access to natural foods, and are allowed to breed freely.

Historically, raising livestock was part of a nomadic or pastoral form of material culture. The herding of camels and reindeer in some parts of the world remains dissociated from sedentary agriculture. The transhumance form of herding in the California Sierra Nevada still continues, as cattle, sheep, or goats are moved from winter pasture in lower-elevation valleys to spring and summer pasture in the foothills and alpine regions, as the seasons progress. Cattle were raised on the open range in the western United States and Canada, on the Pampas of Argentina, and on other prairie and steppe regions of the world.

The enclosure of livestock in pastures and barns is a relatively new development in the history of agriculture. When cattle are enclosed, the type of confinement may vary from a small crate, a large-area fenced-in pasture, or a paddock. The type of feed may vary from naturally growing grass to animal feed. Animals are usually intentionally bred through artificial insemination or supervised mating. Indoor production systems are typically used for pigs, dairy cattle, poultry, veal cattle, dairy goats, and other animals depending on the region and season. Animals kept indoors are generally farmed intensively, as large space requirements could make indoor farming unprofitable if not impossible. However, indoor farming systems are controversial due to problems associated with handling animal waste, odours, the potential for groundwater contamination, and animal welfare concerns. (For a further discussion on intensively farmed livestock, see factory farming, and intensive pig farming). Livestock source verification is used to track livestock.

Other livestock are farmed outdoors, where the size of enclosures and the level of supervision may vary. In large, open ranges, animals may be only occasionally inspected or yarded inround-ups or a muster. Herding dogs may be used for mustering livestock, as are cowboys, stockmen, and jackaroos on horses, in vehicles, and in helicopters. Since the advent of barbed wire (in the 1870s) and electric fence technology, fencing pastures has become much more feasible and pasture management simplified. Rotation of pasturage is a modern technique for improving nutrition and health while avoiding environmental damage to the land. In some cases, very large numbers of animals may be kept in indoor or outdoor feeding operations (on feedlots), where the animals` feed is processed either offsite or onsite, and stored on site before being fed to the animals.

Livestock—-especially cattle—-may be branded to indicate ownership and age, but in modern farming, identification is more likely to be indicated by means of ear tags and electronic identification, instead. Sheep are also frequently marked by means of ear marks and/or ear tags. As fears of BSE and other epidemic illnesses mount, the use of implants to monitor and trace animals in the food production system is increasingly common, and sometimes required by government regulations.

Modern farming techniques seek to minimize human involvement, increase yield, and improve animal health. Economics, quality, and consumer safety all play roles in how animals are raised. The use of hard and soft drugs and feed supplements (or even feed type) may be regulated, or prohibited, to ensure that yield is not increased at the expense of consumer health, safety, or animal welfare. Practices vary around the world, for example growth hormone use is permitted in the United States, but not in stock to be sold in the European Union. The improvement of animal health using modern farming techniques has come into question. Feeding corn to cattle, which have historically eaten grasses, is an example; where the cattle are less adapted to this change, the rumen pH becomes more acidic, leading to liver damage and other health problems. The US Food and Drug Administration allows nonruminant animal proteins to be fed to cattle enclosed in feedlots. For example, it is acceptable to feed chicken manure and poultry meal to cattle, and beef or pork meat and bone meal to chickens.

PREDATION

Livestock farmers have suffered from wild animal predation and theft by rustlers. In North America, animals such as the gray wolf, grizzly bear, cougar, and coyote are sometimes considered a threat to livestock. In Eurasia and Africa, predators include the wolf, leopard, tiger, lion, dhole, Asiatic black bear, crocodile, spotted hyena, and other carnivores. In South America, feral dogs, jaguar, anacondas, and spectacled bears are threats to livestock. In Australia, the dingo, fox, and wedge-tailed eagle are common predators, with an additional threat from domestic dogs that may kill in response to a hunting instinct, leaving the carcass uneaten.

DISEASE

Livestock diseases compromise animal welfare, reduce productivity, and can infect humans. Animal diseases may be tolerated, reduced through animal husbandry, or reduced through antibiotics and vaccines. In developing countries, animal diseases are tolerated in animal husbandry, resulting in considerably reduced productivity, especially given the low health-status of many developing country herds. Disease management to improve productivity is often the first step taken in implementing an agriculture policy.

Disease management can be achieved by modifying animal husbandry practices. These measures aim to prevent infection with biosecurity measures such as controlling animal mixing and entry to farm lots, wearing protective clothing, and quarantining sick animals. Diseases also may be controlled by the use of vaccines and antibiotics. Antibiotics in subtherapeutic doses may also be used as a growth promoter, sometimes increasing growth by 10-15%. Concerns about antibiotic resistance have led in some cases to discouraging the practice of preventive dosing such as the use of antibiotic-laced feed. Countries often require veterinary certificates as a condition for transporting, selling, or exhibiting animals. Disease-free areas often rigorously enforce rules for preventing the entry of potentially diseased animals, including quarantine.

TRANSPORTATION AND MARKETING

Grass-fed cattle, saleyards, Walcha, New South Wales.

Since many livestock are herd animals, they were historically driven to market "on the hoof" to a town or other central location. During the period after the American Civil War, the abundance of Longhorn cattle in Texas and the demand for beef in Northern markets led to the implementation of the Old West cattle drive. The method is still used in some parts of the world. Truck transport is now common in developed countries. Local and regional livestock auctions and commodity markets facilitate trade in livestock. In other areas, livestock may be bought and sold in a bazaar, such as may be found in many parts of Central Asia, or in an informal flea market-type setting.

In developing countries, providing access to markets has encouraged farmers to invest in livestock, with the result being improved livelihoods. The International Crops Research Institute for the Semi-Arid Tropics (ICRISAT) has worked in Zimbabwe to help farmers make their most of their livestock herds. ICRISAT works to improve local farming systems through 'innovation platforms' at which farmers, traders, rural development agencies, and extension officers can discuss the challenges they faced. One finding was that if farmers devoted half of three hectares to maize and half to *mucuna* (velvet bean) in a rotation system, they could obtain 80% of the biomass needed to see their livestock through the dry season. If they only grew maize, they could only meet 20% of their biomass needs. In the town of Gwanda, the platform helped create a strong local market for goats, raising the value of a single animal from US$10 to $60. This gave the farmers a great incentive to invest in their own goats by growing their own feed stock, buying commercial feed only as a supplement, and improving their rangeland management techniques. Because the platform has helped regulate prices, farmers now plan ahead and sell animals at auction, rather than just selling one or two animals at their farm gate as opportunities arise.

Stock shows and fairs are events where people bring their best livestock to compete with one another. Organizations such as 4-H, Block & Bridle, and FFA encourage young people to raise livestock for show purposes. Special feeds are purchased and prior to the show, hours may be spent grooming the animal to look its best. In cattle, sheep, and swine shows, the winning animals are frequently auctioned off to the highest bidder, and the funds are placed into a scholarship fund for its owner. The movie *Grand Champion*, released in 2004, tells the story of a young Texas boy's experience raising a prize steer.

ANIMAL WELFARE

A shepherd boy in India: Livestock are extremely important to the livelihoods of rural smallholder farmers, particularly in the developing world.

The issue of raising livestock for human benefit raises the issue of the suitable relationship between humans and animals, in terms of the status of animals and the obligations of people. The concept of animal welfare reflects the viewpoint that animals under human care should be treated in such a way that they do not suffer unnecessarily. What is considered 'unnecessary' suffering may vary. Generally, though, the animal welfare perspective is based on an interpretation of scientific research on farming practices. By contrast, animal rights defends the viewpoint that using animals for human benefit is, on principle, exploitation, regardless of the farming practices used. Animal rights activists are often vegan or vegetarian, whereas it is consistent with the animal welfare perspective to eat meat as long as the production processes are defensible.

Animal welfare groups generally seek to generate public discussion on livestock raising practices and to secure greater regulation and scrutiny of livestock industry practices. Animal rights groups usually seek to abolish livestock farming, although some groups may recognise the necessity of first achieving more stringent regulation . Animal welfare groups such as the RSPCA are often, in first-world countries, given a voice at governmental level in the development of policy. Animal rights groups find it harder to convey their concerns, and as a result, may advocate civil disobedience or violence.

A number of animal husbandry practices have been the subject of campaigns in the 1990s and 2000s and have led to legislation in some countries. Confinement of livestock in small and unnatural spaces is often done for economic or health reasons. Animals may be kept in the minimum size of cage or pen with little or no space to exercise. Where livestock are used as a source of power, they may be pushed beyond their limits to the point of exhaustion. Increased public awareness and visibility of such abuse meant it was one of the first areas to receive legislation in the 19th century in European countries, but it continues in parts of Asia. Broiler hens may be debeaked, pigs may have deciduous teeth pulled, cattle may be dehorned and branded, dairy cows and sheep may have tails cropped, Merino sheep may undergo mulesing, and many types of male animals may be castrated. Animals may be transported long distances to market and slaughter, often under overcrowded conditions, heat stress, lack of feed and water, and without rest breaks. Such practices have been subject to legislation and protest (see Live Export). Appropriate methods to slaughter livestock were an early target for legislation. Campaigns continue to target halal and kosher religious ritual slaughter.

ENVIRONMENTAL IMPACT

Cattle near the Bruneau River in Elko County, Nevada

Reports such as the United Nations reportLivestock`s Long Shadow cast a pall over the livestock sector (primarily cattle, chickens, and pigs) for `emerging as one of the top two or three most significant contributors to our most serious environmental problems.` In April 2008, the United States Environmental Protection Agency released a major stock-taking of emissions in the United States entitled *Inventory of U.S. Greenhouse Gas Emissions and Sinks: 1990-2006*. It found that "In 2006, the agricultural sector was responsible for emissions of 454.1 teragrams of CO_2 equivalent (Tg CO_2 Eq.), or 6 percent of total U.S. greenhouse gas emissions." By way of comparison, transportation in the US produces more than 25% of all emissions. In 2009, Worldwatch Institute released a report which revealed that 51% of greenhouse gas emissions came from the animal agriculture sector.

The issue of livestock as a major policy focus remains, especially when dealing with problems of deforestation in neotropical areas, land degradation, climate change and air pollution, water shortage and water pollution, and loss of biodiversity. A research team at Obihiro University of Agriculture and Veterinary Medicine in Hokkaidō found that supplementing animals' diets with cysteine, a type of amino acid, and nitrate can reduce the amount of methane gas produced without jeopardising the cattle`s productivity or the quality of their meat and milk.

Deforestation

Deforestation impacts the carbon cycle, as well as the global and regional climate, and causes the habitat loss of many species. Forests that are sinks for the carbon cycle are lost through deforestation. Forests are either logged or burned to make room for mining activities or for grasslands, and often the area needed for such purposes is extensive. Deforestation can also create fragmentation, allowing the survival of only patches of habitat in which species can live. If these patches are distant and small, gene flow is reduced, habitat is altered, edge effects occur, and more opportunities for invasive species to intrude occur.

LAND DEGRADATION

Research from the University of Botswana in 2008 found that farmers' common practice of overstocking cattle to make up for drought losses made ecosystems more vulnerable and risked long-term damage to cattle herds by depleting scarce biomass. The study of the Kgatleng district of Botswana predicted that by 2050, the cycle of mild drought is likely to become shorter for the region (18 months instead of two years) due to climate change.

Climate Change and air Pollution

Testing Australian sheep for exhaled methane production (2001), CSIRO.

Methane is one of the gasses emitted from livestock manure; it persists in the atmosphere for long periods of time and is a potent greenhouse gas, the second-most abundant after carbon dioxide. Though less methane than carbon dioxide is produced, its ability to warm the atmosphere is 25 times greater. Nitrous oxide, another gaseous byproduct of animal agriculture, is about 300 times more potent at trapping heat in the atmosphere. Animal agriculture contributes 65% of anthropogenic nitrous oxide emissions.

Water Shortage

Livestock require water not only for their own consumption, but also for watering the crops needed to produce their feed. Grains are often used to feed livestock; about 50% of US grains and 40% of world grains are used for this purpose Grain and crop production in general require various amounts of water depending on the end product. For example, 100,000 liters of water are needed to yield a kilogram of grain-fed beef, compared to 900 liters for a kilogram of wheat.

Water Pollution

Fertilizers that often contain manure are used to grow crops (such as cereals and fodder) that contain phosphorus and nitrogen, 95% of which is estimated to be lost to the environment. Water pollution from agricultural runoff causes dead zones for plants and aquatic animals due to the lack of oxygen in the water. This lack of oxygen, known as eutrophication, is caused when organisms present in the water grow excessively and then later decompose, in the process using up the oxygen in the water. A prominent example is the Gulf of Mexico, where much of the nutrients in fertilizer used in the US Midwest is funneled down the Mississippi River into the Gulf, causing massive dead zones. Other pollutants not commonly considered are antibiotics and hormones. In southern Asia, vultures that consumed carcasses of livestock declined 95% due to their ingestion of the antibiotic known as diclofenac.

Agricultural Waste

The amount of waste material produced by livestock globally is considerable, as is the effort required to dispose of it. Every minute, 7 million pounds of excrement are produced by animals raised for food in the US (not including animals raised outside of USDA jurisdiction, or aquaculture); another estimate is 335 million tons of "dry matter" / year. A farm with 2,500 dairy cows produces the same amount of waste as a city of 411,000 people. 130 times more animal waste than human waste is produced in the US. – 1.4 billion tons from the meat industry annually. Daily production has been calculated as: dairy cows, 120 lbs x 9 million cows; cattle, 63 lbs x 90 million cattle; pigs, 14 lbs x 67 million pigs; sheep/goats, 5lbs x 9 million sheep/goats; poultry, .25-1 lbs x 9 billion birds.Some waste is disposed of via composting or use as fertilizer. In some countries, animal waste is even used for energy generation. However, improper waste management practices can pollute both surface and groundwater.

Alternatives

Researchers in Australia are looking into the possibility of reducing methane from cattle and sheep by introducing digestive bacteria from kangaroo intestines into livestock. Furthermore, as a means to conserve traditional livestock, cryoconservation of animal genetic resources have been put into action. (Cryoconservation is a practice that involves collecting genetic material and storing it in low temperatures with an intent of conserving a particular breed.)

In semiarid rangelands such as the Great Plains in the U.S., research has provided evidence that livestock can be beneficial to maintaining grassland habitats for big game species.

ECONOMIC AND SOCIAL BENEFITS

The value of global livestock production in 2013 has been estimated at about 883 billion dollars, (constant 2005-2006 dollars). However, economic implications of livestock production extend further: to downstream industry (saleyards, abattoirs, butchers, milk processors, refrigerated transport, wholesalers, retailers, food services, tanneries, etc.), upstream industry (feed producers, feed transport, farm and ranch supply companies, equipment manufacturers, seed companies, vaccine manufacturers, etc.) and associated services (veterinarians, nutrition consultants, shearers, etc.).

Livestock provide a variety of food and nonfood products; the latter include leather, wool, pharmaceuticals, bone products, industrial protein, and fats. For many abattoirs, very little animal biomass may be wasted at slaughter. Even intestinal contents removed at slaughter may be recovered for use as fertilizer. Livestock manure helps maintain the fertility of grazing lands. Manure is commonly collected from barns and feeding areas to fertilize cropland. In some places, animal manure is used as fuel, either directly (as in some developing countries), or indirectly (as a source of methane for heating or for generating electricity). In regions where machine power is limited, some classes of livestock are used as draft stock, not only for tillage and other on-farm use, but also for transport of people and goods. In 1997, livestock provided energy for between an estimated 25 and 64% of cultivation energy in the world's irrigated systems, and that 300 million draft animals were used globally in small-scale agriculture.

Although livestock production serves as a source of income, it can provide additional economic values for rural families, often serving as a major contributor to food security and economic security. Livestock can serve as insurance against risk and is an economic buffer (of income and/or food supply) in some regions and some economies (e.g., during some African droughts). However, its use as a buffer may sometimes be limited where alternatives are present, which may reflect strategic maintenance of insurance in addition to a desire to retain productive assets. Even for some livestock owners in developed nations, livestock can serve as a kind of insurance. Some crop growers may produce livestock as a strategy for diversification of their income sources, to reduce risks related to weather, markets and other factors.

Many studies have found evidence of the social, as well as economic, importance of livestock in developing countries and in regions of rural poverty, and such evidence is not confined to pastoral and nomadic societies.

Social values in developed countries can also be considerable. For example, in a study of livestock ranching permitted on national forest land in New Mexico, USA, it was concluded thatranching maintains traditional values and connects families to ancestral lands and cultural heritage, and that asense of place, attachment to land, and the value of preserving open space were common themes.The importance of land and animals as means of maintaining culture and way of life figured repeatedly in permittee responses, as did the subjects of responsibility and respect for land, animals, family, and community.

In the US, profit tends to rank low among motivations for involvement in livestock ranching. Instead, family, tradition and a desired way of life tend to be major motivators for ranch purchase, and ranchershistorically have been willing to accept low returns from livestock production.

REFERENCES

1. Livestock definition. Dictionary.com. Retrieved 23 November 2015.
2. Cattle | Define Cattle at Dictionary.com. Dictionary.reference.com. Retrieved 2011-12-10.
3. Agriculture: A Glossary of Terms, Programs, and Laws, 2005 Edition (PDF). Archived from the original (PDF) on 2011-02-12. Retrieved 2011-12-10.
4. Breeds of Livestock - Oklahoma State University. Ansi.okstate.edu. Retrieved 2011-12-10.
5. RBST Gene Bank. Rare Breeds Survival Trust. Retrieved 29 December 2015.
6. Northern Daily Leader, 20 May 2010, Dogs mauled 30 sheep (and killed them), p.3, Rural Press
7. Simmons, Michael (2009-09-10).Dogs seized for killing sheep - Local News - News - General - The Times. Victorharbortimes.com.au. Retrieved 2011-12-10.
8. feed (agriculture) :: Antibiotics and other growth stimulants - Britannica Online Encyclopedia. Britannica.com. Retrieved 2011-12-10.
9. Markets from research to outcomes, Farming Matters, Challenge Program on Water and Food, June 2013
10. 2011 U.S. Greenhouse Gas Inventory Report | Climate Change - Greenhouse Gas Emissions | U.S. EPA. Epa.gov. 2006-06-28. Retrieved 2011-12-10.
11. http://www.worldwatch.org/node/6294
12. Global warming breakthrough: way to stop cow gas - Unusual Tales - Specials. Smh.com.au. 2008-01-22. Retrieved 2011-12-10.
13. ‘Livestock`s long shadow: environmental issues and options. Fao.org. Retrieved 2011-12-10.

14. Sub-Saharan Africa news in brief: 10–22 April 2008. SciDev.Net. 2008-04-23. Retrieved 2011-12-10.

15. Global Methane Initiative | Global Methane Initiative. Globalmethane.org. 2010-10-01. Retrieved 2011-12-10.

16. Cornell Science News: Livestock Production. News.cornell.edu. 1997-08-07. Retrieved 2011-12-10.

17. Pelletier Nathan and Peter Tyedmers (2011-01-17).Forecasting Potential Global Environmental Costs of Livestock Production 2000–2050. Proceedings of the National Academy of Sciences of the United States of America 107.43 (2010): 18371-8374. Web of Science.

18. Starme, Elanor (2011-02-05).LEVELING THE FIELD – ISSUE BRIEF #2 Environmental and Health Problems in Livestock Production: Pollution in the Food System (PDF). American Journal of Public Health 94.10: 1703-709.

19. FY-2005 Annual Report Manure and Byproduct Utilization - National Program 206 (PDF). United States Department of Agriculture.

20. Risk assessment evaluation for concentrated animal feeding operations. United States Environmental Protection Agency.

21. Animal agriculture: waste management practices (PDF). United States General Accounting Office.

22. The Facts. COWSPIRACY: The Sustainable Secret.

23. Kangaroo farts could fight global warming | Courier Mail. News.com.au. 2007-12-05. Retrieved 2011-12-10.

24. Derner, Justin D., William K. Lauenroth, Paul Stapp, and David J. Augustine. Livestock as Ecosystem Engineers for Grassland Bird Habitat in the Western Great Plains of North America. Rangeland Ecology & Management 62.2 (2009): 111-18. Web of Science. Web. Feb. 2011.

25. FAOSTAT. (Statistical database of the Food and Agriculture Organization of the United Nations.) http://faostat3.fao.org/

26. de Haan, C. D., H. Steinfeld, and H. Blackburn. 1997. Livestock & the environment: finding a balance. European Commission Directorate-General for Development.

27. Swanepoel, F., A. Stroebel and S. Moyo. (eds.) 2010. The role of livestock in developing communities: Enhancing multifunctionality. African Sun Media.

28. Fafchamps, M., C. Udry, and K. Czukas. 1998. Drought and saving in West Africa: are livestock a buffer stock? Journal of Development Economics 55(2): 273-305

29. Johannesen, A. B. and A. Skonhoft. 2011. Livestock as insurance and social status: Evidence from reindeer herding in Norway. Environmental and Resource Economics, 48(4): 679-694.

30. Bell, L. W. and A. D. Moore. Integrated crop–livestock systems in Australian agriculture: Trends, drivers and implications. Agric. Systems 111: 1-12.

31. Kandulu, J. M., B. A. Bryan, D. King and J. D. Connor. 2012. Mitigating economic risk from climate variability in rain-fed agriculture through enterprise mix diversification. Ecological Economics 79: 105-112.

32. Asresie, A. and L. Zemedu. 2015. Contribution of livestock sector in Ethiopian economy: a review. Advances in Life Science and Technology 29: 79-90.

33. Bettencourt, E. M. V., M. Tilman, P. D. D. S. Henriques, V. Narciso, and M. L. D. S. Carvalho. 2013. The economic and sociocultural role of livestock in the wellbeing of rural communities of Timor-Leste. CEFAGE-UE, Universidade de Évora, Évora, Portugal.

34. Khan, N., A. Rehman and M. Salman. 2013. Impactul cre'terii animalelor asupra dezvoltării socio-economice în Nordul Indiei. Forum geografic 12(1): 75-80.

35. Ali, A. and M. A. Khan. A. 2013. Livestock ownership in ensuring rural household food security in Pakistan. J. Animal Plant Sci. 23(1), 313-318.

36. McSweeney, A. M and C. Raish. 2012. Social, cultural and economic aspects of livestock ranching on the Santa Fe and Carson National Forests. USDA Forest Service RMRS-GTR 276.

37. Gentner, B. J. and J. A. Tanaka. 2002. Classifying federal public land grazing permittees. J. Range Manage. 55 (1): 2-11.

38. Torell, L. A., N. R. Rimbey, J. A. Tanaka, and S. A. Bailey. 2001. The lack of a profit motive for ranching: implications for policy analysis. In: L. A. Torell, E. T. Bartlett and R. Larranaga (eds.) Current issues in rangeland economics. Proc. Symp. Western Coordinating Committee 55. N. M. State Univ. Res. Rep. 737.

CHAPTER 8

Environmental Impact of Agriculture

Chapter Contents

Agriculture, whether conventional or organic, always has some degree of impact on the environment. It is the responsibility of growers and researchers to continuously produce safe and affordable food for the growing population, while minimizing negative impacts on ecosystems and the environment. The environmental impacts of agriculture can be relatively minimal and benign or massive and deleterious.

INTRODUCTION

The **environmental impact of agriculture** varies based on the wide variety of agricultural practices employed around the world. Ultimately, the environmental impact depends on the production practices of the system used by farmers. The connection between emissions into the environment and the farming system is indirect, as it also depends on other climate variables such as rainfall and temperature.

Water pollution **in a rural** stream **due to** runoff from farming activity in New Zealand.

There are two types of indicators of environmental impact: "means-based", which is based on the farmer`s production methods, and "effect-based", which is the impact that farming methods have on the farming system or on emissions to the environment. An example of a means-based indicator would be the quality of groundwater, that is effected by the amount of nitrogen applied to the soil. An indicator reflecting the loss of nitrate to groundwater would be effect-based.

The environmental impact of agriculture involves a variety of factors from the soil, to water, the air, animal and soil diversity, people, plants, and the food itself. Some of the environmental issues that are related to agriculture are climate change, deforestation, genetic engineering, irrigation problems, pollutants, soil degradation, and waste.

ISSUES

Climate Change

Climate change and agriculture are interrelated processes, both of which take place on a worldwide scale. Global warming is projected to have significant impacts on conditions affecting agriculture, including temperature, precipitation and glacial run-off. These conditions determine the carrying capacity of the biosphere to produce enough food for the human population and domesticated animals. Rising carbon dioxide levels would also have effects, both detrimental and beneficial, on crop yields. Assessment of the effects of global climate changes on agriculture might help to properly anticipate and adapt farming to maximize agricultural production. Although the net impact of climate change on agricultural production is uncertain it is likely that it will shift the suitable growing zones for individual crops. Adjustment to this geographical shift will involve considerable economic costs and social impacts.

At the same time, agriculture has been shown to produce significant effects on climate change, primarily through the production and release of greenhouse gases such as carbon dioxide, methane, and nitrous oxide. In addition, agriculture that practices tillage, fertilization, and pesticide application also releases ammonia, nitrate, phosphorus, and many other pesticides that affect air, water, and soil quality, as well as biodiversity. Agriculture also alters the Earth`s land cover, which can change its ability to absorb or reflect heat and light, thus contributing to radiative forcing. Land use change such as deforestation and desertification, together with use of fossil fuels, are the major anthropogenic sources of carbon dioxide; agriculture itself is the major contributor to increasing methane and nitrous oxide concentrations in earth's atmosphere.

DEFORESTATION

Deforestation, clearance or clearing is the removal of a forest or stand of trees where the land is thereafter converted to a non-forest use. Examples of deforestation include conversion of forestland to farms, ranches, or urban use. The most concentrated deforestation occurs in tropical rainforests. About 30% of Earth`s land surface is covered by forests.

In temperate mesic climates, natural regeneration of forest stands often will not occur in the absence of disturbance, whether natural or anthropogenic. Furthermore, biodiversity after regeneration harvest often mimics that found after natural disturbance, including biodiversity loss after naturally occurring rainforest destruction.

Deforestation occurs for multiple reasons: trees are cut down to be used or sold as fuel (sometimes in the form of charcoal) or timber, while cleared land is used as pasture for livestock and plantation. The removal of trees without sufficient reforestation has resulted in damage to habitat, biodiversity loss and aridity. It has adverse impacts on biosequestration of atmospheric carbon dioxide. Deforestation has also been used in war to deprive the enemy of cover for its forces and also vital resources. Modern examples of this were the use of Agent Orange by the British military in Malaya during the Malayan Emergency and the United States military in Vietnam during the Vietnam War. As of 2005, net deforestation rates have ceased to increase in countries with a per capita GDP of at least US$4,600. Deforested regions typically incur significant adverse soil erosion and frequently degrade into wasteland.

Disregard of ascribed value, lax forest management and deficient environmental laws are some of the factors that allow deforestation to occur on a large scale. In many countries, deforestation, both naturally occurring and human-induced, is an ongoing issue. Deforestation causes extinction, changes to climatic conditions, desertification, and displacement of populations as observed by current conditions and in the past through the fossil record. More than half of all plant and land animal species in the world live in tropical forests.

Between 2000 and 2012, 2.3 million square kilometres (890,000 square miles) of forests around the world were cut down. As a result of deforestation, only 6.2 million square kilometres (2.4 million square miles) remain of the original 16 million square kilometres (6 million square miles) of forest that formerly covered the Earth.

Causes

According to the United Nations Framework Convention on Climate Change (UNFCCC) secretariat, the overwhelming direct cause of deforestation is agriculture. Subsistence farming is responsible for 48% of deforestation; commercial agriculture is responsible for 32%; logging is responsible for 14%, and fuel wood removals make up 5%.

Experts do not agree on whether industrial logging is an important contributor to global deforestation. Some argue that poor people are more likely to clear forest because they have no alternatives, others that the poor lack the ability to pay for the materials and labour needed to clear forest. One study found that population increases due to high fertility rates were a primary driver of tropical deforestation in only 8% of cases.

Other causes of contemporary deforestation may include corruption of government institutions, the inequitable distribution of wealth and power, population growth and overpopulation, and urbanization. Globalization is often viewed as another root cause of deforestation, though there are cases in which the impacts of globalization (new flows of labor, capital, commodities, and ideas) have promoted localized forest recovery.

The last batch of sawnwood from the peat forest **in Indragiri Hulu, Sumatra,** Indonesia. Deforestation for oil palm **plantation.**

In 2000 the United Nations Food and Agriculture Organization **(FAO)** found that "the role of population dynamics in a local setting may vary from decisive to negligible," and that deforestation can result from "a combination of population pressure and stagnating economic, social and technological conditions."

The degradation of forest ecosystems has also been traced to economic incentives that make forest conversion appear more profitable than forest conservation. Many important forest functions have no markets, and hence, no economic value that is readily apparent to the forests` owners or the communities that rely on forests for their well-being.

From the perspective of the developing world, the benefits of forest as carbon sinks or biodiversity reserves go primarily to richer developed nations and there is insufficient compensation for these services. Developing countries feel that some countries in the developed world, such as the United States of America, cut down their forests centuries ago and benefited economically from this deforestation, and that it is hypocritical to deny developing countries the same opportunities, i.e. that the poor shouldn`t have to bear the cost of preservation when the rich created the problem.

Some commentators have noted a shift in the drivers of deforestation over the past 30 years. Whereas deforestation was primarily driven by subsistence activities and government-sponsored development projects like transmigration **in countries like** Indonesia **and** colonization **in** Latin America, India, Java, and so on, during the late 19th century and the earlier half of the 20th century, by the 1990s the majority of deforestation was caused by industrial factors, including extractive industries, large-scale cattle ranching, and extensive agriculture.

Problems With Deforestation

Atmospheric

Illegal slash and burn practice in Madagascar, 2010.

Deforestation is ongoing and is shaping climate **and** geography.

Deforestation is a contributor to global warming, and is often cited as one of the major causes of the enhanced greenhouse effect. Tropical deforestation is responsible for approximately 20% of world greenhouse gas emissions.According to the Intergovernmental Panel on Climate Change deforestation, mainly in tropical areas, could account for up to one-third of total anthropogenic carbon dioxide emissions. But recent calculations suggest that carbon dioxide emissions from deforestation and forest degradation (excluding peatland emissions) contribute about 12% of total anthropogenic carbon dioxide emissions with a range from 6 to 17%. Deforestation causes carbon dioxide to linger in the atmosphere. As carbon dioxide accrues, it produces a layer in the atmosphere that traps radiation from the sun. The radiation converts to heat which causes global warming, which is better known as the greenhouse effect. Plants remove carbon in the form of carbon dioxide from the atmosphere during the process of photosynthesis, but release some carbon dioxide back into the atmosphere during normal respiration. Only when actively growing can a tree or forest remove carbon, by storing it in plant tissues. Both the decay and burning of wood releases much of this stored carbon back to the atmosphere. In order for forests to take up carbon, there must be a net accumulation of wood. One way is for the wood to be harvested and turned into long-lived products, with new young trees replacing them. Deforestation may also cause carbon stores held in soil to be released. Forests can be either sinks or sources depending upon environmental circumstances. Mature forests alternate between being net sinks and net sources of carbon dioxide (see carbon dioxide sink and carbon cycle).

In deforested areas, the land heats up faster and reaches a higher temperature, leading to localized upward motions that enhance the formation of clouds and ultimately produce more rainfall. However, according to the Geophysical Fluid Dynamics Laboratory, the models used to investigate remote responses to tropical deforestation showed a broad but mild temperature increase all through the tropical atmosphere. The model predicted <0.2 °C warming for upper air at 700 mb and 500 mb. However, the model shows no significant changes in other areas besides the Tropics. Though the model showed no significant changes to the climate in areas other than the Tropics, this may not be the case since the model has possible errors and the results are never absolutely definite.

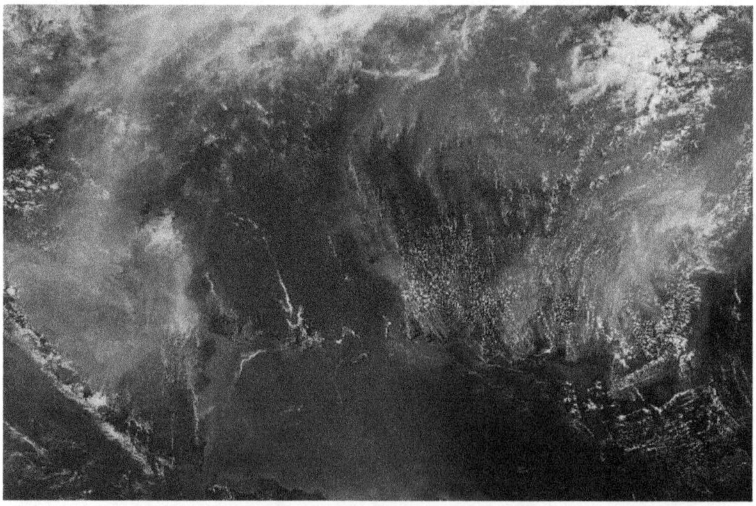

Fires on Borneo and Sumatra, 2006. People use slash-and-burn **deforestation to clear land for agriculture.**

Reducing emissions from deforestation and forest degradation (REDD) in developing countries has emerged as a new potential to complement ongoing climate policies. The idea consists in providing financial compensations for the reduction of greenhouse gas (GHG) emissions from deforestation and forest degradation".

Rainforests are widely believed by laymen to contribute a significant amount of the world`s oxygen, although it is now accepted by scientists that rainforests contribute little net oxygen to the atmosphere and deforestation has only a minor effect on atmospheric oxygen levels. However, the incineration and burning of forest plants to clear land releases large amounts of CO2, which contributes to global warming. Scientists also state that tropical deforestation releases 1.5 billion tons of carbon each year into the atmosphere.

Hydrological

The water cycle is also affected by deforestation. Trees extract groundwater through their roots and release it into the atmosphere. When part of a forest is removed, the trees no longer transpire this water, resulting in a much drier climate. Deforestation reduces the content of water in the soil and groundwater as well as atmospheric moisture. The dry soil leads to lower water intake for the trees to extract. Deforestation reduces soil cohesion, so that erosion, flooding and landslides ensue.

Shrinking forest cover lessens the landscape`s capacity to intercept, retain and transpire precipitation. Instead of trapping precipitation, which then percolates to groundwater systems, deforested areas become sources of surface water runoff, which moves much faster than subsurface flows. Forests return most of the water that falls as precipitation to the atmosphere by transpiration. In contrast, when an area is deforested, almost all precipitation is lost as run-off. That quicker transport of surface water can translate into flash flooding and more localized floods than would occur with the forest cover. Deforestation also contributes to decreased evapotranspiration, which lessens atmospheric moisture which in some cases affects precipitation levels downwind from the deforested area, as water is not recycled to downwind forests, but is lost in runoff and returns directly to the oceans. According to one study, in deforested north and northwest China, the average annual precipitation decreased by one third between the 1950s and the 1980s.

Trees, and plants in general, affect the water cycle **significantly:**

- their canopies intercept a proportion of precipitation, which is then evaporated back to the atmosphere (canopy interception);
- their litter, stems and trunks slow down surface runoff;
- their roots create macropores – large conduits – in the soil that increase infiltration of water;
- they contribute to terrestrial evaporation and reduce soil moisture via transpiration;
- their litter and other organic residue change soil properties that affect the capacity of soil to store water.
- their leaves control the humidity of the atmosphere by transpiring. 99% of the water absorbed by the roots moves up to the leaves and is transpired.

As a result, the presence or absence of trees can change the quantity of water on the surface, in the soil or groundwater, or in the atmosphere. This in turn changes erosion rates and the availability of water for either ecosystem functions or human services.

The forest may have little impact on flooding in the case of large rainfall events, which overwhelm the storage capacity of forest soil if the soils are at or close to saturation.

Tropical rainforests **produce about 30% of our planet's** fresh water.

Soil

Deforestation for the use of clay **in the** Brazilian **city of** Rio de Janeiro. The hill depicted is Morro da Covanca, in Jacarepaguá.

Undisturbed forests have a very low rate of soil loss (erosion), approximately 2 metric tons per square kilometer (6 short tons per square mile). Deforestation generally increases rates of soil loss, by increasing the amount of runoff and reducing the protection of the soil from tree litter. This can be an advantage in excessively leached tropical rain forest soils. Forestry operations themselves also increase erosion through the development of (forest) roads and the use of mechanized equipment.

China`s Loess Plateau was cleared of forest millennia ago. Since then it has been eroding, creating dramatic incised valleys, and providing the sediment that gives the Yellow River its yellow color and that causes the flooding of the river in the lower reaches (hence the river's nickname 'China's sorrow').

Removal of trees does not always increase erosion rates. In certain regions of southwest US, shrubs and trees have been encroaching on grassland. The trees themselves enhance the loss of grass between tree canopies. The bare intercanopy areas become highly erodible. The US Forest Service, in Bandelier National Monument for example, is studying how to restore the former ecosystem, and reduce erosion, by removing the trees.

Tree roots bind soil together, and if the soil is sufficiently shallow they act to keep the soil in place by also binding with underlying bedrock. Tree removal on steep slopes with shallow soil thus increases the risk of landslides, which can threaten people living nearby.

Biodiversity

Deforestation on a human scale results in decline in biodiversity, and on a natural global scale is known to cause the extinction of many species.{ The removal or destruction of areas of forest cover has resulted in a degraded environment with reduced biodiversity. Forests support biodiversity, providing habitat for wildlife; moreover, forests foster medicinal conservation. With forest biotopes being irreplaceable source of new drugs (such as taxol), deforestation can destroy genetic variations (such as crop resistance) irretrievably.

Illegal logging in Madagascar. In 2009, the vast majority of the illegally obtained rosewood **was exported to** China.

Since the tropical rainforests are the most diverse ecosystems on Earth and about 80% of the world's known biodiversity could be found in tropical rainforests, removal or destruction of significant areas of forest cover has resulted in a degraded environment with reduced biodiversity. A study in Rondônia, Brazil, has shown that deforestation also removes the microbial community which is involved in the recycling of nutrients, the production of clean water and the removal of pollutants.

It has been estimated that we are losing 137 plant, animal and insect species every single day due to rainforest deforestation, which equates to 50,000 species a year. Others state that tropical rainforest deforestation is contributing to the ongoing Holocene mass extinction. The known extinction rates from deforestation rates are very low, approximately 1 species per year from mammals and birds which extrapolates to approximately 23,000 species per year for all species. Predictions have been made that more than 40% of the animal and plant species in Southeast Asia could be wiped out in the 21st century. Such predictions were called into question by 1995 data that show that within regions of Southeast

Asia much of the original forest has been converted to monospecific plantations, but that potentially endangered species are few and tree flora remains widespread and stable.

Scientific understanding of the process of extinction is insufficient to accurately make predictions about the impact of deforestation on biodiversity. Most predictions of forestry related biodiversity loss are based on species-area models, with an underlying assumption that as the forest declines species diversity will decline similarly. However, many such models have been proven to be wrong and loss of habitat does not necessarily lead to large scale loss of species.Species-area models are known to overpredict the number of species known to be threatened in areas where actual deforestation is ongoing, and greatly overpredict the number of threatened species that are widespread.

A recent study of the Brazilian Amazon predicts that despite a lack of extinctions thus far, up to 90 percent of predicted extinctions will finally occur in the next 40 years.

Economic Impact

Damage to forests and other aspects of nature could halve living standards for the world's poor and reduce global GDP by about 7% by 2050, a report concluded at the Convention on Biological Diversity (CBD) meeting in Bonn in 2008. Historically, utilization of forest products, including timber and fuel wood, has played a key role in human societies, comparable to the roles of water and cultivable land. Today, developed countries continue to utilize timber for building houses, and wood pulp for paper. In developing countries almost three billion people rely on wood for heating and cooking.

The forest products industry is a large part of the economy in both developed and developing countries. Short-term economic gains made by conversion of forest to agriculture, or over-exploitation of wood products, typically leads to loss of long-term income and long-term biological productivity. West Africa, Madagascar, Southeast Asia and many other regions have experienced lower revenue because of declining timber harvests. Illegal logging causes billions of dollars of losses to national economies annually.

The new procedures to get amounts of wood are causing more harm to the economy and overpower the amount of money spent by people employed in logging.According to a study, "in most areas studied, the various ventures that prompted deforestation rarely generated more than US$5 for every ton of carbon they released and frequently returned far less than US$1". The price on the European market for an offset tied to a one-ton reduction in carbon is 23 euro (about US$35).

Rapidly growing economies also have an effect on deforestation. Most pressure will come from the world`s developing countries, which have the fastest-growing populations and most rapid economic (industrial) growth. In 1995, economic growth in developing countries reached nearly 6%, compared with the 2% growth rate for developed countries." As our human population grows, new homes, communities, and expansions of cities will occur. Connecting all of the new expansions will be roads, a very important part in our daily life. Rural roads promote economic development but also facilitate deforestation. About 90% of the deforestation has occurred within 100 km of roads in most parts of the Amazon.

The European Union is one of the largest importer of products made from illegal deforestation.

Forest Transition Theory

The forest transition and historical baselines.

The forest area change may follow a pattern suggested by the forest transition **(FT)** theory, whereby at early stages in its development a country is characterized by high forest cover and low deforestation rates (HFLD countries).

Then deforestation rates accelerate (HFHD, high forest cover – high deforestation rate), and forest cover is reduced (LFHD, low forest cover – high deforestation rate), before the deforestation rate slows (LFLD, low forest cover – low deforestation rate), after which forest cover stabilizes and eventually starts recovering. FT is not a "law of nature," and the pattern is influenced by national context (for example, human population density, stage of development, structure of the economy), global economic forces, and government policies. A country may reach very low levels of forest cover before it stabilizes, or it might through good policies be able to "bridge" the forest transition.

FT depicts a broad trend, and an extrapolation of historical rates therefore tends to underestimate future BAU deforestation for counties at the early stages in the transition (HFLD), while it tends to overestimate BAU deforestation for countries at the later stages (LFHD and LFLD).

Countries with high forest cover can be expected to be at early stages of the FT. GDP per capita captures the stage in a country's economic development, which is linked to the pattern of natural resource use, including forests. The choice of forest cover and GDP per capita also fits well with the two key scenarios in the FT:

(i) a forest scarcity path, where forest scarcity triggers forces (for example, higher prices of forest products) that lead to forest cover stabilization; and

(ii) an economic development path, where new and better off-farm employment opportunities associated with economic growth (= increasing GDP per capita) reduce profitability of frontier agriculture and slows deforestation.

Historical causes

Prehistory

The Carboniferous Rainforest Collapse was an event that occurred 300 million years ago. Climate change devastated tropical rainforests causing the extinction of many plant and animal species. The change was abrupt, specifically, at this time climate became cooler and drier, conditions that are not favourable to the growth of rainforests and much of the biodiversity within them. Rainforests were fragmented forming shrinking `islands` further and further apart. This sudden collapse affected several large groups, effects on amphibians were particularly devastating, while reptiles fared better, being ecologically adapted to the drier conditions that followed.

An array of Neolithic artifacts, including bracelets, axe heads, chisels, and polishing tools.

Rainforests once covered 14% of the earth's land surface; now they cover a mere 6% and experts estimate that the last remaining rainforests could be consumed in less than 40 years. Small scale deforestation was practiced by some societies for tens of thousands of years before the beginnings of civilization. The first evidence of deforestation appears in the Mesolithic period. It was probably used to convert closed forests into more open ecosystems favourable to game animals. With the advent of agriculture, larger areas began to be deforested, and fire became the prime tool to clear land for crops. In Europe there is little solid evidence before 7000 BC. Mesolithic foragers used fire to create openings for red deer and wild boar. In Great Britain, shade-tolerant species such as oak and ash are replaced in the pollen record by hazels, brambles, grasses and nettles. Removal of the forests led to decreased transpiration, resulting in the formation of upland peat bogs. Widespread decrease in elm pollen across Europe between 8400–8300 BC and 7200–7000 BC, starting in southern Europe and gradually moving north to Great Britain, may represent land clearing by fire at the onset of Neolithic agriculture.

The Neolithic period saw extensive deforestation for farming land. Stone axes were being made from about 3000 BC not just from flint, but from a wide variety of hard rocks from across Britain and North America as well. They include the noted Langdale axe industry in the English Lake District, quarries developed at Penmaenmawr in North Wales and numerous other locations. Rough-outs were made locally near the quarries, and some were polished locally to give a fine finish. This step not only increased the mechanical strength of the axe, but also made penetration of wood easier. Flint was still used from sources such as Grimes Graves but from many other mines across Europe.

Evidence of deforestation has been found in Minoan Crete; for example the environs of the Palace of Knossos were severely deforested in the Bronze Age.

Pre-Industrial History

Easter Island, deforested. According to Jared Diamond: “Among past societies faced with the prospect of ruinous deforestation, Easter Island and Mangareva chiefs succumbed to their immediate concerns, but Tokugawa shoguns, Inca emperors, New Guinea highlanders, and 16th century German landowners adopted a long view and reafforested.”

Throughout most of history, humans were hunter gatherers who hunted within forests. In most areas, such as the Amazon, the tropics, Central America, and the Caribbean, only after shortages of wood and other forest products occur are policies implemented to ensure forest resources are used in a sustainable manner.

In ancient Greece, Tjeered van Andel and co-writers summarized three regional studies of historic erosion and alluviation and found that, wherever adequate evidence exists, a major phase of erosion follows, by about 500-1,000 years the introduction of farming in the various regions of Greece, ranging from the later Neolithic to the Early Bronze Age. The thousand years following the mid-first millennium BC saw serious, intermittent pulses of soil erosion in numerous places. The historic silting of ports along the southern coasts of Asia Minor (e.g. Clarus, and the examples of Ephesus, Priene and

Miletus, where harbors had to be abandoned because of the silt deposited by the Meander) and in coastal Syria during the last centuries BC.

Easter Island has suffered from heavy soil erosion in recent centuries, aggravated by agriculture and deforestation. Jared Diamond gives an extensive look into the collapse of the ancient Easter Islanders in his book Collapse. The disappearance of the island`s trees seems to coincide with a decline of its civilization around the 17th and 18th century. He attributed the collapse to deforestation and over-exploitation of all resources.

The famous silting up of the harbor for Bruges, which moved port commerce to Antwerp, also followed a period of increased settlement growth (and apparently of deforestation) in the upper river basins. In early medieval Riez in upper Provence, alluvial silt from two small rivers raised the riverbeds and widened the floodplain, which slowly buried the Roman settlement in alluvium and gradually moved new construction to higher ground; concurrently the headwater valleys above Riez were being opened to pasturage.

A typical progress trap was that cities were often built in a forested area, which would provide wood for some industry (for example, construction, shipbuilding, pottery). When deforestation occurs without proper replanting, however; local wood supplies become difficult to obtain near enough to remain competitive, leading to the city's abandonment, as happened repeatedly in Ancient Asia Minor. Because of fuel needs, mining and metallurgy often led to deforestation and city abandonment.

With most of the population remaining active in (or indirectly dependent on) the agricultural sector, the main pressure in most areas remained land clearing for crop and cattle farming. Enough wild green was usually left standing (and partially used, for example, to collect firewood, timber and fruits, or to graze pigs) for wildlife to remain viable. The elite`s (nobility and higher clergy) protection of their own hunting privileges and game often protected significant woodlands.

Major parts in the spread (and thus more durable growth) of the population were played by monastical `pioneering` (especially by the Benedictine and Commercial orders) and some feudal lords' recruiting farmers to settle (and become tax payers) by offering relatively good legal and fiscal conditions. Even when speculators sought to encourage towns, settlers needed an agricultural belt around or sometimes within defensive walls. When populations were quickly decreased by causes such as the Black Death or devastating warfare (for example, Genghis Khan`s Mongol hordes in eastern and central Europe, Thirty Years` War in Germany), this could lead to settlements being abandoned. The land was reclaimed by nature, but the secondary forests usually lacked the original biodiversity.

From 1100 to 1500 AD, significant deforestation took place in Western Europe as a result of the expanding human population. The large-scale building of wooden sailing ships by European (coastal) naval owners since the 15th century for exploration, colonisation, slave trade–and other trade on the high seas consumed many forest resources. Piracy also contributed to the over harvesting of forests, as in Spain. This led to a weakening of the domestic economy after Columbus' discovery of America, as the economy became dependent on colonial activities (plundering, mining, cattle, plantations, trade, etc.)

In Changes in the Land (1983), William Cronon analyzed and documented 17th-century English colonists' reports of increased seasonal flooding in New England during the period when new settlers initially cleared the forests for agriculture. They believed flooding was linked to widespread forest clearing upstream.

Deforestation of Brazil's Atlantic Forest **c.1820-1825.**

The massive use of charcoal on an industrial scale in Early Modern Europe was a new type of consumption of western forests; even in Stuart England, the relatively primitive production of charcoal has already reached an impressive level. Stuart England was so widely deforested that it depended on the Baltic trade for ship timbers, and looked to the untapped forests of New England to supply the need. Each of Nelson's Royal Navy war ships at Trafalgar (1805) required 6,000 mature oaks for its construction. In France, Colbert planted oak forests to supply the French navy in the future. When the oak plantations matured in the mid-19th century, the masts were no longer required because shipping had changed.

Norman F. Cantor`s summary of the effects of late medieval deforestation applies equally well to Early Modern Europe:

Europeans had lived in the midst of vast forests throughout the earlier medieval centuries. After 1250 they became so skilled at deforestation that by 1500 they were running short of wood for heating and cooking. They were faced with a nutritional decline because of the elimination of the generous supply of wild game that had inhabited the now-disappearing forests, which throughout medieval times had provided the staple of their carnivorous high-protein diet. By 1500 Europe was on the edge of a fuel and nutritional disaster [from] which it was saved in the sixteenth century only by the burning of soft coal and the cultivation of potatoes and maize.

Industrial Era

In the 19th century, introduction of steamboats in the United States was the cause of deforestation of banks of major rivers, such as the Mississippi River, with increased and more severe flooding one of the environmental results. The steamboat crews cut wood every day from the riverbanks to fuel the steam engines. Between St. Louis and the confluence with the Ohio River to the south, the Mississippi became more wide and shallow, and changed its channel laterally. Attempts to improve navigation by the use of snag pullers often resulted in crews' clearing large trees 100 to 200 feet (61 m) back from the banks. Several French colonial towns of the Illinois Country, such as Kaskaskia, Cahokia and St. Philippe, Illinois were flooded and abandoned in the late 19th century, with a loss to the cultural record of their archeology.

The wholescale clearance of woodland to create agricultural land can be seen in many parts of the world, such as the Central forest-grasslands transition and other areas of the Great Plains of the United States. Specific parallels are seen in the 20th-century deforestation occurring in many developing nations.

Rates of Deforestation

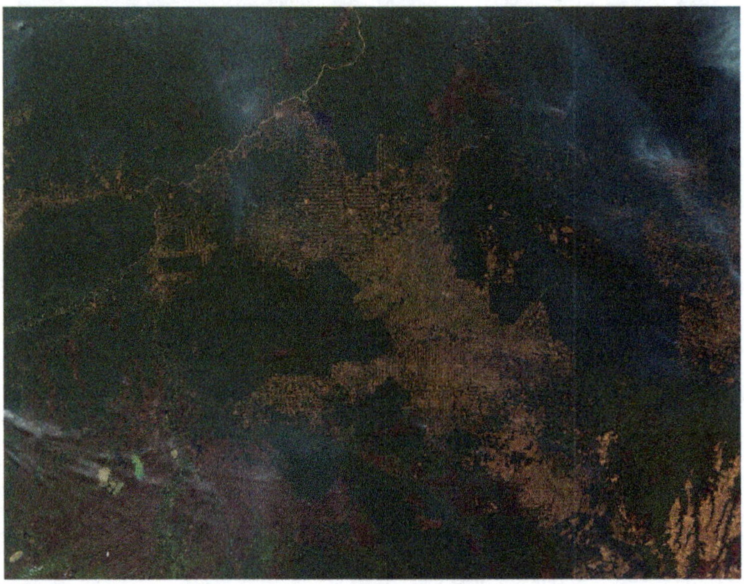

Slash-and-burn **farming in the state of** Rondônia, western Brazil.

Global deforestation sharply accelerated around 1852. It has been estimated that about half of the Earth`s mature tropical forests—between 7.5 million and 8 million km2 (2.9 million to 3 million sq mi) of the original 15 million to 16 million km2 (5.8 million to 6.2 million sq mi) that until 1947 covered the planet—have now been destroyed. Some scientists have predicted that unless significant measures (such as seeking out and protecting old growth forests that have not been disturbed) are taken on a worldwide basis, by 2030 there will only be 10% remaining, with another 10% in a degraded condition. 80% will have been lost, and with them hundreds of thousands of irreplaceable species. Some cartographers have attempted to illustrate the sheer scale of deforestation by country using a cartogram.

Estimates vary widely as to the extent of tropical deforestation. Scientists estimate that one fifth of the world`s tropical rainforest was destroyed between 1960 and 1990. They claim that that rainforests 60 years ago covered 14% of the world`s land surface, now only cover 5–7%, and that all tropical forests will be gone by the middle of the 21st century.

A 2002 analysis of satellite imagery suggested that the rate of deforestation in the humid tropics (approximately 5.8 million hectares per year) was roughly 23% lower than the most commonly quoted rates. Conversely, a newer analysis of satellite images reveals that deforestation of the Amazon rainforest is twice as fast as scientists previously estimated.

Some have argued that deforestation trends may follow a Kuznets curve, which if true would nonetheless fail to eliminate the risk of irreversible loss of non-economic forest values (for example, the extinction of species).

Satellite image of Haiti's border with the Dominican Republic (right) shows the amount of deforestation on the Haitian side.

A 2005 report by the United Nations Food and Agriculture Organization (FAO) estimates that although the Earth`s total forest area continues to decrease at about 13 million hectares per year, the global rate of deforestation has recently been slowing. Still others claim that rainforests are being destroyed at an ever-quickening pace. The London-based Rainforest Foundation notes that "the UN figure is based on a definition of forest as being an area with as little as 10% actual tree cover, which would therefore include areas that are actually savannah-like ecosystems and badly damaged forests."Other critics of the FAO data point out that they do not distinguish between forest types, and that they are based largely on reporting from forestry departments of individual countries, which do not take into account unofficial activities like illegal logging.

Despite these uncertainties, there is agreement that destruction of rainforests remains a significant environmental problem. Up to 90% of West Africa`s coastal rainforests have disappeared since 1900. In South Asia, about 88% of the rainforests have been lost. Much of what remains of the world`s rainforests is in the Amazon basin, where the Amazon

Rainforest covers approximately 4 million square kilometres. The regions with the highest tropical deforestation rate between 2000 and 2005 were Central America—which lost 1.3% of its forests each year—and tropical Asia. In Central America, two-thirds of lowland tropical forests have been turned into pasture since 1950 and 40% of all the rainforests have been lost in the last 40 years. Brazil has lost 90–95% of its Mata Atlântica forest. Paraguay was losing its natural semi humid forests in the country's western regions at a rate of 15.000 hectares at a randomly studied 2-month period in 2010, Paraguay's parliament refused in 2009 to pass a law that would have stopped cutting of natural forests altogether.

Deforestation around Pakke Tiger Reserve, India.

Madagascar has lost 90% of its eastern rainforests. As of 2007, less than 50% of Haiti`s forests remained. Mexico, India, the Philippines, Indonesia, Thailand, Burma, Malaysia, Bangladesh, China, Sri Lanka, Laos, Nigeria, the Democratic Republic of the Congo, Liberia, Guinea, Ghana and the Ivory Coast, have lost large areas of their rainforest. Several countries, notably Brazil, have declared their deforestation a national emergency. The World Wildlife Fund`s ecoregion project catalogues habitat types throughout the world, including habitat loss such as deforestation, showing for example that even in the rich forests of parts of Canada such as the Mid-Continental Canadian forests of the prairie provinces half of the forest cover has been lost or altered.

Regions

Rates of deforestation vary around the world.

In 2011 Conservation International listed the top 10 most endangered forests, characterized by having all lost 90% or more of their original habitat, and each harboring at least 1500 endemic plant species (species found nowhere else in the world).

Top 10 Most Endangered Forests 2011				
Endangered forest	**Region**	**Remaining habitat**	**Predominate vegetation type**	**Notes**
Indo-Burma	Asia-Pacific	5%	Tropical and subtropical moist broadleaf forests	Rivers, floodplain wetlands, mangrove forests. Burma, Thailand, Laos, Vietnam, Cambodia, India.
New Caledonia	Asia-Pacific	5%	Tropical and subtropical moist broadleaf forests	See note for region covered.
Sundaland	Asia-Pacific	7%	Tropical and subtropical moist broadleaf forests	Western half of the Indo-Malayan archipelago including southern Borneo and Sumatra.
Philippines	Asia-Pacific	7%	Tropical and subtropical moist broadleaf forests	Forests over the entire country including 7,100 islands.
Atlantic Forest	South America	8%	Tropical and subtropical moist broadleaf forests	Forests along Brazil's Atlantic coast, extends to parts of Paraguay, Argentina and Uruguay.
Mountains of Southwest China	Asia-Pacific	8%	Temperate coniferous forest	See note for region covered.
California Floristic Province	North America	10%	Tropical and subtropical dry broadleaf forests	See note for region covered.
Coastal Forests of Eastern Africa	Africa	10%	Tropical and subtropical moist broadleaf forests	Mozambique, Tanzania, Kenya, Somalia.

Madagascar & Indian Ocean Islands	Africa	10%	Tropical and subtropical moist broadleaf forests	Madagascar, Mauritius, Reunion, Seychelles, Comoros.
Eastern Afromontane	Africa	11%	Tropical and subtropical moist broadleaf forests Montane grasslands and shrublands	Forests scattered along the eastern edge of Africa, from Saudi Arabia in the north to Zimbabwe in the south.

Control

Reducing Emissions

Main international organizations including the United Nations and the World Bank, have begun to develop programs aimed at curbing deforestation. The blanket term Reducing Emissions from Deforestation and Forest Degradation (REDD) describes these sorts of programs, which use direct monetary or other incentives to encourage developing countries to limit and/or roll back deforestation. Funding has been an issue, but at the UN Framework Convention on Climate Change (UNFCCC) Conference of the Parties-15 (COP-15) in Copenhagen in December 2009, an accord was reached with a collective commitment by developed countries for new and additional resources, including forestry and investments through international institutions, that will approach USD 30 billion for the period 2010–2012. Significant work is underway on tools for use in monitoring developing country adherence to their agreed REDD targets. These tools, which rely on remote forest monitoring using satellite imagery and other data sources, include the Center for Global Development`s FORMA (Forest Monitoring for Action) initiative and the Group on Earth Observations` Forest Carbon Tracking Portal. Methodological guidance for forest monitoring was also emphasized at COP-15. The environmental organization Avoided Deforestation Partners leads the campaign for development of REDD through funding from the U.S. government. In 2014, the Food and Agriculture Organization of the United Nations and partners launched Open Foris - a set of open-source software tools that assist countries in gathering, producing and disseminating information on the state of forest resources. The tools support the inventory lifecycle, from needs assessment, design, planning, field data collection and management, estimation analysis, and dissemination. Remote sensing image processing tools are included, as well as tools for international reporting for Reducing emissions from deforestation and forest degradation (REDD) and MRV and FAO`s Global Forest Resource Assessments.

In evaluating implications of overall emissions reductions, countries of greatest concern are those categorized as High Forest Cover with High Rates of Deforestation

(HFHD) and Low Forest Cover with High Rates of Deforestation (LFHD). Afghanistan, Benin, Botswana, Burma, Burundi, Cameroon, Chad, Ecuador, El Salvador, Ethiopia, Ghana, Guatemala, Guinea, Haiti, Honduras, Indonesia, Liberia, Malawi, Mali, Mauritania, Mongolia, Namibia, Nepal, Nicaragua, Niger, Nigeria, Pakistan, Paraguay, Philippines, Senegal, Sierra Leone, Sri Lanka, Sudan, Togo, Uganda, United Republic of Tanzania, Zimbabwe are listed as having Low Forest Cover with High Rates of Deforestation (LFHD). Brazil, Cambodia, Democratic Peoples Republic of Korea, Equatorial Guinea, Malaysia, Solomon Islands, Timor-Leste, Venezuela, Zambia are listed as High Forest Cover with High Rates of Deforestation (HFHD).

Payments for Conserving Forests

In Bolivia, deforestation in upper river basins has caused environmental problems, including soil erosion and declining water quality. An innovative project to try and remedy this situation involves landholders in upstream areas being paid by downstream water users to conserve forests. The landholders receive US$20 to conserve the trees, avoid polluting livestock practices, and enhance the biodiversity and forest carbon on their land. They also receive US$30, which purchases a beehive, to compensate for conservation for two hectares of water-sustaining forest for five years. Honey revenue per hectare of forest is US$5 per year, so within five years, the landholder has sold US$50 of honey. The project is being conducted by Fundación Natura Bolivia **and Rare Conservation, with support from the** Climate & Development Knowledge Network.

Land rights

Transferring land rights **to indigenous inhabitants is argued to efficiently conserve forests.**

Transferring rights over land from public domain to its indigenous inhabitants is argued to be a cost effective strategy to conserve forests. This includes the protection of such rights entitled in existing laws, such as India's Forest Rights Act. The transferring of such rights in China, perhaps the largest land reform in modern times, has been argued to have increased forest cover. In Brazil, forested areas given tenure to indigenous groups have even lower rates of clearing than national parks.

Farming

New methods are being developed to farm more intensively, such as high-yield hybrid crops, greenhouse, autonomous building gardens, and hydroponics. These methods are often dependent on chemical inputs to maintain necessary yields. In cyclic agriculture, cattle are grazed on farm land that is resting and rejuvenating. Cyclic agriculture actually increases the fertility of the soil. Intensive farming can also decrease soil nutrients by consuming at an accelerated rate the trace minerals needed for crop growth.The most promising approach, however, is the concept of food forests in permaculture, which consists of agroforestal systems carefully designed to mimic natural forests, with an emphasis on plant and animal species of interest for food, timber and other uses. These systems have low dependence on fossil fuels and agro-chemicals, are highly self-maintaining, highly productive, and with strong positive impact on soil and water quality, and biodiversity.

Monitoring Deforestation

There are multiple methods that are appropriate and reliable for reducing and monitoring deforestation. One method is the "visual interpretation of aerial photos or satellite imagery that is labor-intensive but does not require high-level training in computer image processing or extensive computational resources". Another method includes hot-spot analysis (that is, locations of rapid change) using expert opinion or coarse resolution satellite data to identify locations for detailed digital analysis with high resolution satellite images. Deforestation is typically assessed by quantifying the amount of area deforested, measured at the present time. From an environmental point of view, quantifying the damage and its possible consequences is a more important task, while conservation efforts are more focused on forested land protection and development of land-use alternatives to avoid continued deforestation. Deforestation rate and total area deforested, have been widely used for monitoring deforestation in many regions, including the Brazilian Amazon deforestation monitoring by INPE. A global satellite view is available.

Forest Management

Efforts to stop or slow deforestation have been attempted for many centuries because it has long been known that deforestation can cause environmental damage sufficient in some cases to cause societies to collapse. In Tonga, paramount rulers developed policies designed to prevent conflicts between short-term gains from converting forest to farmland and long-term problems forest loss would cause, while during the 17th and 18th centuries in Tokugawa, Japan, the shoguns developed a highly sophisticated system of long-term planning to stop and even reverse deforestation of the preceding centuries through substituting timber by other products and more efficient use of land that had been farmed for many centuries. In 16th-century Germany, landowners also developed silviculture to deal with the problem of deforestation. However, these policies tend to be limited to environments with good rainfall, no dry season and very young soils (through volcanism or glaciation). This is because on older and less fertile soils trees grow too slowly for silviculture to be economic, whilst in areas with a strong dry season there is always a risk of forest fires destroying a tree crop before it matures.

In the areas where "slash-and-burn" is practiced, switching to "slash-and-char" would prevent the rapid deforestation and subsequent degradation of soils. The biochar thus created, given back to the soil, is not only a durable carbon sequestration method, but it also is an extremely beneficial amendment to the soil. Mixed with biomass it brings the creation of terra preta, one of the richest soils on the planet and the only one known to regenerate itself.

Sustainable Practices

Bamboo is advocated as a more sustainable alternative for cutting down wood for fuel.

Certification, as provided by global certification systems such as Programme for the Endorsement of Forest Certification and Forest Stewardship Council, contributes to tackling deforestation by creating market demand for timber from sustainably managed forests. According to the United Nations Food and Agriculture Organization (FAO), "A major condition for the adoption of sustainable forest management is a demand for products that are produced sustainably and consumer willingness to pay for the higher costs entailed. Certification represents a shift from regulatory approaches to market incentives to promote sustainable forest management. By promoting the positive attributes of forest products from sustainably managed forests, certification focuses on the demand side of environmental conservation." Rainforest Rescue argues that the standards of organizations like FSC are too closely connected to timber industry interests and therefore do not guarantee environmentally and socially responsible forest management. In reality, monitoring systems are inadequate and various cases of fraud have been documented worldwide.

Some nations have taken steps to help increase the amount of trees on Earth. In 1981, China created National Tree Planting Day Forest and forest coverage had now reached 16.55% of China`s land mass, as against only 12% two decades ago

Using fuel from bamboo rather than wood results in cleaner burning, and since bamboo matures much faster than wood, deforestation is reduced as supply can be replenished faster.

Reforestation

In many parts of the world, especially in East Asian countries, reforestation and afforestation are increasing the area of forested lands. The amount of woodland has increased in 22 of the world's 50 most forested nations. Asia as a whole gained 1 million hectares of forest between 2000 and 2005. Tropical forest in El Salvador expanded more than 20% between 1992 and 2001. Based on these trends, one study projects that global forest will increase by 10%—an area the size of India—by 2050.

In the People`s Republic of China, where large scale destruction of forests has occurred, the government has in the past required that every able-bodied citizen between the ages of 11 and 60 plant three to five trees per year or do the equivalent amount of work in other forest services. The government claims that at least 1 billion trees have been planted in China every year since 1982. This is no longer required today, but March 12 of every year in China is the Planting Holiday. Also, it has introduced the Green Wall of China project, which aims to halt the expansion of the Gobi desert through the planting of trees. However, due to the large percentage of trees dying off after planting (up to 75%), the project is not very successful. There has been a 47-million-hectare increase in forest area in China since the 1970s. The total number of trees amounted to be about 35 billion and 4.55% of China's land mass increased in forest coverage. The forest coverage was 12% two decades ago and now is 16.55%.

An ambitious proposal for China is the Aerially Delivered Re-forestation and Erosion Control System and the proposed Sahara Forest Project coupled with the Seawater Greenhouse.

In Western countries, increasing consumer demand for wood products that have been produced and harvested in a sustainable manner is causing forest landowners and forest industries to become increasingly accountable for their forest management and timber harvesting practices.

The Arbor Day Foundation`s Rain Forest Rescue program is a charity that helps to prevent deforestation. The charity uses donated money to buy up and preserve rainforest land before the lumber companies can buy it. The Arbor Day Foundation then protects the land from deforestation. This also locks in the way of life of the primitive tribes living on the forest land. Organizations such as Community Forestry International, Cool Earth, The Nature Conservancy, World Wide Fund for Nature, Conservation International, African Conservation Foundation and Greenpeace also focus on preserving forest habitats. Greenpeace in particular has also mapped out the forests that are still intact and published this information on the internet. World Resources Institute in turn has made a simpler thematic map showing the amount of forests present just before the age of man (8000 years ago) and the current (reduced) levels of forest. These maps mark the amount of afforestation required to repair the damage caused by people.

Forest Plantations

To meet the world's demand for wood, it has been suggested by forestry writers Botkins and Sedjo that high-yielding forest plantations are suitable. It has been calculated that plantations yielding 10 cubic meters per hectare annually could supply all the timber required for international trade on 5% of the world's existing forestland. By contrast, natural forests produce about 1–2 cubic meters per hectare; therefore, 5–10 times more forestland would be required to meet demand. Forester Chad Oliver has suggested a forest mosaic with high-yield forest lands interspersed with conservation land.

In the country of Senegal, on the western coast of Africa, a movement headed by youths has helped to plant over 6 million mangrove trees. The trees will protect local villages from storm damages and will provide a habitat for local wildlife. The project started in 2008, and already the Senegalese government has been asked to establish rules and regulations that would protect the new mangrove forests.

Military Context

While the preponderance of deforestation is due to demands for agricultural and urban use for the human population, there are some examples of military causes. One example of deliberate deforestation is that which took place in the U.S. zone of occupation in Germany after World War II. Before the onset of the Cold War, defeated Germany was still considered a potential future threat rather than potential future ally. To address this threat, attempts were made to lower German industrial potential, of which forests were deemed an element. Sources in the U.S. government admitted that the purpose of this was that the "ultimate destruction of the war potential of German forests." As a consequence of the practice of clear-felling, deforestation resulted which could "be replaced only by long forestry development over perhaps a century."

American Sherman tanks knocked out by Japanese artillery on Okinawa.

Deforestation can also be one consequence of war. For example, in the 1945 Battle of Okinawa, bombardment and other combat operations reduced the lush tropical landscape into "a vast field of mud, lead, decay and maggots".Deforestation can also be an intentional tactic of military forces. Defoliants (like Agent Orange or others) was used by the British in the Malayan Emergency, and by the United States in the Korean War and Vietnam War.

Public Health Context

Deforestation eliminates a great number of species of plants and animals which also often results in an increase in disease. Loss of native species allows new species to come to dominance. Often the destruction of predatory species can result in an increase in rodent populations. These are known to carry plagues. Additionally, erosion can produce pools of stagnant water that are perfect breeding grounds for mosquitos, well known vectors of malaria, yellow fever, nipah virus, and more. Deforestation can also create a path for non-native species to flourish such as certain types of snails, which have been correlated with an increase in schistosomiasis cases.

Deforestation is occurring all over the world and has been coupled with an increase in the occurrence of disease outbreaks. In Malaysia, thousands of acres of forest have been cleared for pig farms. This has resulted in an increase in the zoonosis the Nipah virus. In Kenya, deforestation has led to an increase in malaria cases which is now the leading cause of morbidity and mortality the country.

Another pathway through which deforestation affects disease is the relocation and dispersion of disease-carrying hosts. This disease emergence pathway can be called "range expansion," whereby the host's range (and thereby the range of pathogens) expands to new geographic areas. Through deforestation, hosts and reservoir species are forced into neighboring habitats. Accompanying the reservoir species are pathogens that have the ability to find new hosts in previously unexposed regions. As these pathogens and species come into closer contact with humans, they are infected both directly and indirectly.

A catastrophic example of range expansion is the 1998 outbreak of Nipah Virus in Malaysia. For a number of years, deforestation, drought, and subsequent fires led to a dramatic geographic shift and density of fruit bats, a reservoir for Nipah virus. Deforestation reduced the available fruiting trees in the bats' habitat, and they encroached on surrounding orchards which also happened to be the location of a large number of pigsties. The bats, through proximity spread the Nipah to pigs. While the virus infected the pigs, mortality was much lower than among humans, making the pigs a virulent host leading to the transmission of the virus to humans. This resulted in 265 reported cases of encephalitis, of which 105 resulted in death. This example provides an important lesson for the impact deforestation can have on human health.

Another example of range expansion due to deforestation and other anthropogenic habitat impacts includes the Capybara rodent in Paraguay. This rodent is the host of a number of zoonotic diseases and, while there has not yet been a human-borne outbreak due to the movement of this rodent into new regions, it offers an example of how habitat destruction through deforestation and subsequent movements of species is occurring regularly.

A now well-developed theory is that the spread of HIV it is at least partially due deforestation. Rising populations created a food demand and with deforestation opening up new areas of the forest the hunters harvested a great deal of primate bushmeat, which is believed to be the origin of HIV.

Genetic Engineering

Genetically engineered crops are herbicide-tolerant, and their overuse has created herbicide resistant “super weeds”,which may ultimately increase the use of herbicides. Seed contamination is another problem of genetic engineering; it can occur from wind or bee pollination that is blown from genetically-engineered crops to normal crops. About 50% of corn and soybean samples and more than 80% of canola samples were found to be contaminated by Monsanto`s (genetic engineering company) genes.This accidental contamination can cause organic farmers to lose a lot of money because they need to recall their products. There are various cases of this such as in the corn and alfalfa industry.

Irrigation

Irrigation can lead to a number of problems:

Among some of these problems is the depletion of underground aquifers through overdrafting. Soil can be over-irrigated because of poor distribution uniformity or management wastes water, chemicals, and may lead to water pollution. Over-irrigation can cause deep drainage from rising water tables that can lead to problems of irrigation salinity requiring watertable control by some form of subsurface land drainage. However, if the soil is under irrigated, it gives poor soil salinity control which leads to increased soil salinity with consequent buildup of toxic salts on soil surface in areas with high evaporation. This requires either leaching to remove these salts and a method of drainage to carry the salts away. Irrigation with saline or high-sodium water may damage soil structure owing to the formation of alkaline soil.

Pollutants

Synthetic pesticides such as ‘Malathon, ‘Rogor’, ‘Kelthane’ and ‘confidor’ are the most widespread method of controlling pests in agriculture. Pesticides can leach through the soil and enter the groundwater, as well as linger in food products and result in death in humans. Pesticides can also kill non-target plants, birds, fish and other wildlife. A wide range of agricultural chemicals are used and some become pollutants through use, misuse, or ignorance. Pollutants from agriculture have a huge effect on water quality. Agricultural nonpoint source (NPS) solution impacts lakes, rivers, wetlands, estuaries, and groundwater. Agricultural NPS can be caused by poorly managed animal feeding operations, overgrazing, plowing, fertilizer, and improper, excessive, or badly timed use of Pesticides. Pollutants from farming include sediments, nutrients, pathogens, pesticides, metals, and salts.

Listed below are additional and specific problems that may arise with the release of pollutants from agriculture.

- Pesticide drift
- soil contamination
- air pollution *spray drift*
- Pesticides, especially those based on organochloride
- Pesticide residue in foods
- Pesticide toxicity to bees
- List of crop plants pollinated by bees
- Pollination management
- Bioremediation

Soil Degradation

Soil degradation is the decline in soil quality that can be a result of many factors, especially from agriculture. Soils hold the majority of the world's biodiversity, and healthy soils are essential for food production and an adequate water supply. Common attributes of soil degradation can be salting, waterlogging, compaction, pesticide contamination, decline in soil structure quality, loss of fertility, changes in soil acidity, alkalinity, salinity, and erosion. Soil degradation also has a huge impact on biological degradation, which affects the microbial community of the soil and can alter nutrient cycling, pest and disease control, and chemical transformation properties of the soil.

- soil contamination
- sedimentation

Waste

Plasticulture is the use of plastic mulch in agriculture. Farmers use plastic sheets as mulch to cover 50-70% of the soil and allows them to use drip irrigation systems to have better control over soil nutrients and moisture. Rain is not required in this system, and farms that use plasticulture are built to encourage the fastest runoff of rain. The use of pesticides with plasticulture allows pesticides to be transported easier in the surface runoff towards wetlands or tidal creeks. The runoff from pesticides and chemicals in the plastic can cause serious deformations and death in shellfish as the runoff carries the chemicals towards the oceans.

In addition to the increased runoff that results from plasticulture, there is also the problem of the increased amount of waste form the plastic mulch itself. The use of plastic mulch for vegetables, strawberries, and other row and orchard crops exceeds 110 million pounds annually in the United States. Most plastic ends up in the landfill, although there are other disposal options such as disking mulches into the soil, on-site burying, on-site storage, reuse, recycling, and incineration. The incineration and recycling options are complicated by the variety of the types of plastics that are used and by the geographic dispersal of the plastics. Plastics also contain stabilizers and dyes as well as heavy metals, which limits the amount of products that can be recycled. Research is continually being conducted on creating biodegradable or photodegradable mulches. While there has been minor success with this, there is also the problem of how long the plastic takes to degrade, as many biodegradable products take a long time to break down.

Issues by Region

The environmental impact of agriculture can vary depending on the region as well as the type of agriculture production method that is being used. Listed below are some specific environmental issues in a various different regions around the world.

- Hedgerow removal in the United Kingdom.
- Soil salinisation, especially in Australia.
- Phosphate mining in Nauru
- Methane emissions from livestock in New Zealand. *See Climate change in New Zealand.*
- Environmentalists attribute the hypoxic zone in the Gulf of Mexico as being encouraged by nitrogen fertilization of the algae bloom.

SUSTAINABLE AGRICULTURE

The exponential population increase in recent decades has increased the practice of agricultural land conversion to meet demand for food which in turn has increased the effects on the environment. The global population is still increasing and will eventually stabilise, as some critics doubt that food production, due to lower yields from global warming, can support the global population.

Organic farming is a multifaceted sustainable agriculture set of practices that can have a lower impact on the environment at the small scale. However in most cases organic farming results in lower yields in terms of production per unit area and per unit of irrigation water. Therefore, widespread adoption of organic agriculture will require additional land to be cleared and water resources extracted to meet the same level of production. A European meta-analysis found that organic farms tended to have higher soil organic matter content and lower nutrient losses (nitrogen leaching, nitrous oxide emissions and ammonia emissions) per unit of field area but higher ammonia emissions, nitrogen leaching and nitrous oxide emissions per product unit.

Other specific methods include: permaculture; and biodynamic agriculture which incorporates a spiritual element.

- Category: Sustainable agriculture
- Biological pest control

REFERENCES

1. "Un dizième des terres sauvages ont disparu en deux décennies" (Radio Télévision Suisse) citing Watson, James E.M.; Shanahan, Danielle F.; Di Marco, Moreno; Allan, James; Laurance, William F.; Sanderson, Eric W.; MacKey, Brendan;

Venter, Oscar (2016). "Catastrophic Declines in Wilderness Areas Undermine Global Environment Targets". Current Biology. 26 (21): 2929–2934. doi:10.1016/j.cub.2016.08.049. PMID 27618267.

2. SAFnet Dictionary|Definition For [deforestation]. Dictionary of forestry.org (2008-07-29). Retrieved 2011-05-15.
3. Bradford, Alina. (2015-03-04) Deforestation: Facts, Causes & Effects. Livescience.com. Retrieved 2016-11-13.
4. Deforestation | Threats | WWF. Worldwildlife.org. Retrieved 2016-11-13.
5. Oliver, C.D. (1980). "Forest Development in North America following major disturbances". For. Ecol. Management. 3: 153–168. doi:10.1016/0378-1127(80)90013-4.
6. Sahney, S., Benton, M.J. & Falcon-Lang, H.J. (2010). "Rainforest collapse triggered Pennsylvanian tetrapod diversification in Euramerica". Geology. 38 (12): 1079–1082. doi:10.1130/G31182.1.
7. Patel-Weynand, Toral (2002) Biodiversity and sustainable forestry: State of the science review. The National Commission on Science for Sustainable Forestry, Washington DC
8. Kauppi, P. E.; Ausubel, J. H.; Fang, J.; Mather, A. S.; Sedjo, R. A.; Waggoner, P. E. (2006). "Returning forests analyzed with the forest identity". Proceedings of the National Academy of Sciences. 103 (46): 17574–9. doi:10.1073/pnas.0608343103. PMC 1635979 . PMID 17101996.
9. "Use Energy, Get Rich and Save the Planet", The New York Times, 20 April 2009.
10. Rainforest Facts. Nature.org (2016-11-01). Retrieved 2016-11-13.
11. "Facts About Rainforests". The Nature Conservancy. Retrieved 2015-10-19.
12. UNFCCC (2007). "Investment and financial flows to address climate change" (PDF). unfccc.int. UNFCCC. p. 81.
13. Arild Angelsen; David Kaimowitz (February 1999). "Rethinking the causes of deforestation: Lessons from economic models". The World Bank Research Observer, 14:1. Oxford University Press. pp. 73–98.
14. Laurance, William F. (December 1999). "Reflections on the tropical deforestation crisis" (PDF). Biological Conservation, Volume 91, Issues 2–3. pp. 109–117.
15. Helmut J. Geist And Eric F. Lambin (February 2002). "Proximate Causes and Underlying Driving Forces of Tropical Deforestation" (PDF). BioScience. 52 (2): 143–150. doi:10.1641/0006-3568(2002)052[0143:PCAUDF]2.0.CO;2.
16. Burgonio, T.J. (3 January 2008). "Corruption blamed for deforestation". Philippine Daily Inquirer.
17. "WRM Bulletin Number 74". World Rainforest Movement. September 2003.
18. "Global Deforestation". Global Change Curriculum. University of Michigan Global Change Program. 4 January 2006.
19. Alain Marcoux (August 2000). "Population and deforestation". SD Dimensions.

Sustainable Development Department, Food and Agriculture Organization of the United Nations (FAO).

20. Butler, Rhett A. "Impact of Population and Poverty on Rainforests". Mongabay.com / A Place Out of Time: Tropical Rainforests and the Perils They Face. Retrieved 13 May 2009.
21. Stock, Jocelyn; Rochen, Andy. "The Choice: Doomsday or Arbor Day". umich.edu.
22. Karen. "Demographics, Democracy, Development, Disparity and Deforestation: A Crossnational Assessment of the Social Causes of Deforestation". Paper presented at the annual meeting of the American Sociological Association, Atlanta Hilton Hotel, Atlanta, GA, 16 August 2003. Retrieved 13 May 2009.
23. "The Double Edge of Globalization". YaleGlobal Online. Yale University Press. June 2007.
24. Butler, Rhett A. "Human Threats to Rainforests—Economic Restructuring". Mongabay.com / A Place Out of Time: Tropical Rainforests and the Perils They Face. Retrieved 13 May 2009.
25. Susanna B. Hecht; Susan Kandel; Ileana Gomes; Nelson Cuellar; Herman Rosa (2006). "Globalization, Forest Resurgence, and Environmental Politics in El Salvador"(PDF). World Development. 34 (2): 308–323. doi:10.1016/j.worlddev.2005.09.005.
26. Pearce, David W (December 2001). "The Economic Value of Forest Ecosystems" (PDF). Ecosystem Health. 7 (4): 284–296. doi:10.1046/j.1526-0992.2001.01037.x.
27. Erwin H Bulte; Mark Joenje; Hans G P Jansen (2000). "Is there too much or too little natural forest in the Atlantic Zone of Costa Rica?". Canadian Journal of Forest Research. 30 (3): 495–506. doi:10.1139/x99-225.
28. Butler, Rhett A.; Laurance, William F. (August 2008). "New strategies for conserving tropical forests" (PDF). Trends in Ecology & Evolution. 23 (9): 469–472. doi:10.1016/j.tree.2008.05.006.
29. Rudel, T.K. 2005 "Tropical Forests: Regional Paths of Destruction and Regeneration in the Late 20th Century"Columbia University Press ISBN 0-231-13195-X
30. "NASA – Top Story – NASA DATA SHOWS DEFORESTATION AFFECTS CLIMATE".
31. "Massive deforestation threatens food security".
32. Deforestation, ScienceDaily
33. Confirmed: Deforestation Plays Critical Climate Change Role, ScienceDaily, 11 May 2007.
34. Clearing Forests May Transform Local—and Global—Climate; Researchers are finding that massive deforestation may have a profound, and possibly catastrophic, impact on local weather 4 March 2013 Scientific American
35. Deforestation causes global warming, FAO
36. Philip M. Fearnside1 and William F. Laurance, TROPICAL DEFORESTATION

AND GREENHOUSE-GAS EMISSIONS, Ecological Applications, Volume 14, Issue 4 (August 2004) pp. 982–986.

37. "Fondation Chirac " Deforestation and desertification".
38. IPCC Fourth Assessment Report, Working Group I Report "The Physical Science Basis", Section 7.3.3.1.5 (p. 527)
39. G.R. van der Werf, D.C.Morton, R.S. DeFries, J.G.J. Olivier, P.S. Kasibhatla, R.B. Jackson, G.J. Collatz and J.T. Randerson (2009). "CO2 emissions from forest loss". Nature Geoscience. 2 (11): 737–738. doi:10.1038/ngeo671.
40. Mumoki, Fiona. "The Effects of Deforestation on our Environment Today." Panorama. TakingITGlobal. 18 July 2006. Web. 24 March 2012.
41. I.C. Prentice. "The Carbon Cycle and Atmospheric Carbon Dioxide" IPCC
42. NASA Data Shows Deforestation Affects Climate In The Amazon. NASA News. 9 June 2004.
43. Findell, Kristen L.; Thomas R. Knutson (2006). "Weak Simulated Extratropical Responses to Complete Tropical Deforestation". Journal of Climate. 19 (12): 2835–2850. doi:10.1175/JCLI3737.1.
44. Sheila Wertz-Kanounnikoff; Laura Ximena Rubio Alvarado. "Why are we seeing "REDD"?". Institute for Sustainable Development and International Relations. Archived from the original on 25 December 2007. Retrieved 2016-11-14.
45. "How can you save the rain forest. 8 October 2006. Frank Field". The Times. London. 8 October 2006. Retrieved 1 April 2010.
46. Broeker, Wallace S. (2006). "Breathing easy: Et tu, O2."Columbia University
47. Moran, Emilio F. (1993). "Deforestation and land use in the Brazilian Amazon". Human Ecology. 21: 1–21. doi:10.1007/BF00890069.
48. R Defries; F Achard; S Brown; M Herold; D Murdiyarso; B Schlamadinger; C Desouzajr (2007). "Earth observations for estimating greenhouse gas emissions from deforestation in developing countries" (PDF). Environmental Science Policy. 10 (4): 385–394. doi:10.1016/j.envsci.2007.01.010.
49. "Underlying Causes of Deforestation". UN Secretary-General's Report.
50. Daniel Rogge. "Deforestation and Landslides in Southwestern Washington". University of Wisconsin-Eau Claire.
51. China's floods: Is deforestation to blame? BBC News. 6 August 1999.
52. The book "Environment", 5th Edition, Raven and Berg, John Wiley & Sons, 2006, page 406.
53. Hongchang, Wang (1998-01-01). "Deforestation and Desiccation in China A Preliminary Study". In Schwartz, Jonathan Matthew. The Economic Costs of China's Environmental Degradation: Project on Environmental Scarcities, State Capacity, and Civil Violence, a Joint Project of the University of Toronto and the American Academy of Arts and Sciences. Committee on Internat. Security Studies, American Acad. of Arts and Sciences.
54. "Soil, Water and Plant Characteristics Important to Irrigation". North Dakota

State University.

55. Sten Nilsson, Do We Have Enough Forests?, American Institute of Biological Sciences, March 2001.
56. Rainforest Biodiversity Shows Differing Patterns, ScienceDaily, 14 August 2007.
57. "Medicine from the rainforest". Research for Biodiversity Editorial Office. Archived from the original on 6 December 2008.
58. Single-largest biodiversity survey says primary rainforest is irreplaceable, Bio-Medicine, 14 November 2007.
59. Tropical rainforests – The tropical rainforest, BBC
60. Tropical Rain Forest. thinkquest.org
61. U.N. calls on Asian nations to end deforestation, Reuters, 20 June 2008.
62. "Rainforest Facts".
63. Tropical rainforests – Rainforest water and nutrient cycles, BBC
64. Butler, Rhett A. (2 July 2007) Primary rainforest richer in species than plantations, secondary forests, mongabay.com,
65. Flowers, April. "Deforestation In The Amazon Affects Microbial Life As Well As Ecosystems". Science News. Redorbit.com. Retrieved 12 March 2013.
66. Rainforest Facts. Rain-tree.com (2010-03-20). Retrieved 2010-08-29.
67. Leakey, Richard and Roger Lewin, 1996, The Sixth Extinction : Patterns of Life and the Future of Humankind, Anchor, ISBN 0-385-46809-1.
68. The great rainforest tragedy, The Independent, 28 June 2003.
69. Biodiversity wipeout facing South East Asia, New Scientist, 23 July 2003.
70. Pimm, S. L.; Russell, G. J.; Gittleman, J. L.; Brooks, T. M. (1995). "The Future of Biodiversity". Science. 269 (5222): 347–350. doi:10.1126/science.269.5222.347. PMID 17841251.
71. Pimm Stuart L, Russell Gareth J, Gittleman John L, Brooks Thomas M (1995). "The future of biodiversity". Science. 269 (5222): 347–341. doi:10.1126/science.269.5222.347. PMID 17841251.
72. Timothy Charles Whitmore; Jeffrey Sayer; International Union for Conservation of Nature and Natural Resources. General Assembly; IUCN Forest Conservation Programme (15 February 1992). Tropical deforestation and species extinction. Springer. ISBN 978-0-412-45520-9. Retrieved 4 December 2011.
73. Sohn, Emily (12 July 2012). "More extinctions expected in Amazon". Discovery.
74. Nature loss 'to hurt global poor', BBC News, 29 May 2008.
75. Forest Products. (PDF). Retrieved 2011-12-04.
76. "Destruction of Renewable Resources". rainforests.mongabay.com.
77. Deforestation Across the World's Tropical Forests Emits Large Amounts of Greenhouse Gases with Little Economic Benefits, According to a New Study at CGIAR.org, 4 December 2007.

78. "New ASB Report finds deforestation offers very little money compared to potential financial benefits". asb.cgiar.org.
79. Chomitz, Kenneth; Gray, David A. (1999). "Roads, lands, markets, and deforestation : a spatial model of land use in Belize". Policy Research Working Papers. doi:10.1596/1813-9450-1444.
80. Ferraz, Silvio Frosini de Barros; Vettorazzi, Carlos Alberto; Theobald, David M. (2009). "Using indicators of deforestation and land-use dynamics to support conservation strategies: A case study of central Rondônia, Brazil". Forest Ecology and Management. 257 (7): 1586–1595. doi:10.1016/j.foreco.2009.01.013.
81. "Stolen Goods: The EU's complicity in illegal tropical deforestation" (PDF). Forests and the European Union Resource Network. 17 March 2015. Retrieved 31 March 2015.
82. "Reducing Emissions from Deforestation and Forest Degradation (REDD): An Options Assessment Report"(PDF). The Government of Norway. 28 March 2010. p. 16.
83. Meyfroidt, P., Lambin, E.F. 2011. Global Forest Transition: Prospects for an End to Deforestation. Annual Review of Environment and Resources 36: 343-371.
84. Taylor, Leslie (2004). The Healing Power of Rainforest Herbs: A Guide to Understanding and Using Herbal Medicinals. Square One. ISBN 9780757001444.
85. Flannery, T (1994). The future eaters. Melbourne: Reed Books. ISBN 0-7301-0422-2.
86. Brown, Tony (1997). "Clearances and Clearings: Deforestation in Mesolithic/Neolithic Britain". Oxford Journal of Archaeology. 16 (2): 133–146. doi:10.1111/1468-0092.00030.
87. "hand tool: Neolithic tools". Encyclopædia Britannica Online.
88. "Neolithic Age from 4,000 BC to 2,200 BC or New Stone Age".
89. Hogan, C. Michael (22 December 2007). "Knossos fieldnotes", The Modern Antiquarian
90. Jared Diamond, Collapse: How Societies Choose to Fail or Survive, chapter "The world as a polder: what does it all mean to us today?", section "Reasons for hope", Penguin Books, 2005 and 2011, pages 522–523 (ISBN 978-0-241-95868-1).
91. Encyclopćdia Britannica Online School Edition. School.eb.com. Retrieved 2010-08-29.
92. Tjeerd H. van Andel; Eberhard Zangger; Anne Demitrack (2013). "Land Use and Soil Erosion in Prehistoric and Historical Greece" (PDF). Journal of Field Archaeology. 17 (4): 379–396. doi:10.1179/009346990791548628.
93. "The Mystery of Easter Island", Smithsonian Magazine, 1 April 2007.
94. "Historical Consequences of Deforestation: Easter Island".
95. "Jared Diamond, Easter Island's End".
96. Chew, Sing C. (2001). World Ecological Degradation. Oxford, England: AltaMira

Press. pp. 69–70.

97. Norman F. Cantor (9 June 1994). The civilization of the Middle Ages: a completely revised and expanded edition of Medieval history, the life and death of a civilization. HarperCollins. p. 564. ISBN 978-0-06-092553-6. Retrieved 4 December 2011.
98. F. Terry Norris, "Where Did the Villages Go? Steamboats, Deforestation, and Archaeological Loss in the Mississippi Valley", in Common Fields: an environmental history of St. Louis, Andrew Hurley, ed., St. Louis, MO: Missouri Historical Society Press, 1997, ISBN 978-1-883982-15-7 pp. 73–89.
99. Duke Press policy studies / Global deforestation and the nineteenth-century world economy / edited by Richard P. Tucker and J. F. Richards
100. E. O. Wilson, 2002, The Future of Life, Vintage ISBN 0-679-76811-4.
101. Map reveals extent of deforestation in tropical countries, guardian.co.uk, 1 July 2008.
102. Maycock, Paul F. Deforestation. WorldBookOnline.
103. Ron Nielsen, The Little Green Handbook: Seven Trends Shaping the Future of Our Planet, Picador, New York (2006) ISBN 978-0-312-42581-4.
104. Victor Vescovo. (2006). The Atlas of World Statistics. The Caladan Press. Retrieved 2012-08-03.
105. Teja Tscharntke, Christoph Leuschner, Edzo Veldkamp, Heiko Faust, Edi Guhardja, eds. (2010). Tropical Rainforests and Agroforests Under Global Change. Springer. pp. 270–271. ISBN 978-3-642-00492-6.
106. Intergovernmental Panel on Climate Change (2000). Land Use, Land Use Change and Forestry. Cambridge University Press.
107. John F. Mongillo; Linda Zierdt-Warshaw (2000). Linda Zierdt-Warshaw, ed. Encyclopedia of environmental science. University of Rochester Press. p. 104. ISBN 978-1-57356-147-1.
108. Achard Frederic, Eva Hugh D, Hans- , Stibig Jurgen, Mayaux Philippe (2002). "Determination of deforestation rates of the world's humid tropical forests". Science. 297 (5583): 999–1003. doi:10.1126/science.1070656. PMID 12169731.
109. Jha, Alok (21 October 2005). "Amazon rainforest vanishing at twice rate of previous estimates". The Guardian.
110. Satellite images reveal Amazon forest shrinking faster, csmonitor.com, 21 October 2005.
111. Culas, Richard J. (2007). "Deforestation and the environmental Kuznets curve: An institutional perspective"(PDF). Ecological Economics. 61 (2–3): 429–437. doi:10.1016/j.ecolecon.2006.03.014.
112. Environmental Economics: A deforestation Kuznets curve?, 22 November 2006.
113. "Is there an environmental Kuznets curve for deforestation?".
114. "Pan-tropical Survey of Forest Cover Changes 1980–2000". Forest Resources

Assessment. Rome, Italy: Food and Agriculture Organization of the United Nations (FAO).

115. Committee On Forestry. FAO (2001-03-16). Retrieved 2010-08-29.
116. Worldwatch: Wood Production and Deforestation Increase & Recent Content, Worldwatch Institute
117. Rhett A. Butler (16 November 2005). "World deforestation rates and forest cover statistics, 2000–2005". mongabay.com.
118. The fear is that highly diverse habitats, such as tropical rainforest, are vanishing at a faster rate that is partly masked by the slower deforestation of less biodiverse, dry, open forests. Because of this omission, the most harmful impacts of deforestation (such as habitat loss) could be increasing despite a possible decline in the global rate of deforestation.
119. "Remote sensing versus self-reporting".
120. The World Bank estimates that 80% of logging operations are illegal in Bolivia and 42% in Colombia, while in Peru, illegal logging accounts for 80% of all logging activities. (World Bank (2004). Forest Law Enforcement.) (The Peruvian Environmental Law Society (2003). Case Study on the Development and Implementation of Guidelines for the Control of Illegal Logging with a View to Sustainable Forest Management in Peru.)
121. "Forest Holocaust". National Geographic.
122. "Rainforests & Agriculture".
123. The Amazon Rainforest, BBC, 14 February 2003.
124. John Revington. "The Causes of Tropical Deforestation". New Renaissance Magazine.
125. "What is Deforestation?". kids.mongabay.com.
126. "Paraguay es principal deforestador del Chaco". ABC Color newspaper, Paraguay. Retrieved 13 August 2011.
127. "Paraguay farmland". Retrieved 13 August 2011.
128. IUCN – Three new sites inscribed on World Heritage List, 27 June 2007.
129. "Madagascar's rainforest map". New Scientist.
130. "Haiti Is Covered with Trees". EnviroSociety. Tarter, Andrew. Retrieved 14 November 2016.
131. Chart – Tropical Deforestation by Country & Region. Mongabay.com. Retrieved 2011-12-04.
132. Rainforest Destruction. rainforestweb.org
133. Amazon deforestation rises sharply in 2007, USATODAY.com, 24 January 2008.
134. Vidal, John (31 May 2005). "Rainforest loss shocks Brazil". The Guardian. London. Retrieved 1 April 2010.
135. The World's 10 Most Threatened Forest Hotspots, Conservation International, 2 February 2011.

136. Indo-Burma, Conservation International.

137. New Caledonia, Conservation International.

138. Sundaland, Conservation International.

139. Philippines, Conservation International.

140. Atlantic Forest Archived 12 December 2011 at the Wayback Machine., Conservation International.

141. Mountains of Southwest China, Conservation International.

142. California Floristic Province Archived 14 April 2011 at the Wayback Machine., Conservation International.

143. Coastal Forests of Eastern Africa, Conservation International.

144. Madagascar & Indian Ocean Islands, Conservation International.

145. Eastern Afromontane, Conservation International.

146. "Copenhagen Accord of 18 December 2009" (PDF). UNFCC. 2009. Retrieved 2009-12-28.

147. Forest Monitoring for Action (FORMA) : Center for Global Development : Initiatives: Active. Cgdev.org (2009-11-23). Retrieved 2010-08-29.

148. Browser – GEO FCT Portal. Portal.geo-fct.org. Retrieved 2010-08-29.

149. "Methodological Guidance" (PDF). UNFCC. 2009. Retrieved 2009-12-28.

150. Agriculture Secretary Vilsack: $1 billion for REDD+ " Climate Progress. Climateprogress.org (2009-12-16). Retrieved 2010-08-29.

151. "FAO sets standards to improve national forest monitoring systems".

152. Angelsen, Arild; et al. (2009). "Reducing Emissions from Deforestation and Forest Degradation (REDD): An Options Assessment Report" (PDF). Meridian Institute for the Government of Norway. pp. 75–77. Retrieved 2011-11-24.

153. Payments for watershed services: A driver of climate compatible development, Climate & Development Knowledge Network, 30 December 2013.

154. "India should follow China to find a way out of the woods on saving forest people". The Guardian. 22 July 2016. Retrieved 7 August 2016.

155. "China's forest tenure reforms". rightsandresources.org. Retrieved 7 August 2016.

156. "Global Forest Change - Google Crisis Map". Google Crisis Map. Retrieved 12 October 2016.

157. "Warning to forest destroyers: this scientist will catch you". Nature News & Comment. 4 October 2016. Retrieved 12 October 2016.

158. Diamond, Jared Collapse: How Societies Choose To Fail or Succeed; Viking Press 2004, pp. 301–302 ISBN 0-14-311700-9.

159. Diamond, Jared Collapse: How Societies Choose To Fail or Succeed; Viking Press 2004, pp. 320–331 ISBN 0-14-311700-9.

160. Rosenberg, Tina (2012-03-13). "In Africa's vanishing forests, the benefits of bamboo". New York Times. Retrieved 2012-07-26.

161. "State of the World's Forests 2009". United Nations Food and Agriculture Organization.
162. Facts about Tropical Timber. Rainforest Rescue. Retrieved 2016-11-13.
163. John Gittings. "Battling China's Deforestation." The Guardian. 20 March 2001.
164. Foley Jonathan A; DeFries Ruth; Asner Gregory P; Barford Carol; et al. (2005). "Global Consequences of Land Use". Science. 309 (5734): 570–574. doi:10.1126/science.1111772. PMID 16040698.
165. James Owen, "World's Forests Rebounding, Study Suggests" National Geographic News, 13 November 2006.
166. John Gittings, "Battling China's deforestation", World News, 20 March 2001.
167. The world's last intact forest landscapes. intactforests.org
168. "World Intact Forests campaign by Greenpeace". intactforests.org.
169. The World's Forests from a Restoration Perspective, WRI
170. "Alternative thematic map by Howstuffworks; in pdf"(PDF).
171. Daniel B. Botkin (2001). No man's garden: Thoreau and a new vision for civilization and nature. Island Press. pp. 246–247. ISBN 978-1-55963-465-6. Retrieved 4 December 2011.
172. Stenstrup, Allen (2010). Forests. Greensboro, North Carolina: Morgan Reynolds Publishing. p. 89. ISBN 978-1-59935-116-2.
173. Balabkins, Nicholas (1964) "Germany Under Direct Controls; Economic Aspects Of Industrial Disarmament 1945–1948, Rutgers University Press. p. 119. The two quotes used by Balabkins are referenced to, respectively: U.S. office of Military Government, A Year of Potsdam: The German Economy Since the Surrender (1946), p. 70; and U.S. Office of Military Government, The German Forest Resources Survey (1948), p. II. For similar observations see G.W. Harmssen, Reparationen, Sozialproduct, Lebensstandard (Bremen: F. Trujen Verlag, 1948), I, 48.
174. Higa, Takejiro. Battle of Okinawa, The Hawaii Nisei Project
175. Pesticide Dilemma in the Third World: A Case Study of Malaysia. Phoenix Press. 1984. p. 23.
176. "Encyclopedia of World Environmental History". Routledge, 2004. ISBN 0-415-93733-7.
177. Marchak, M. Patricia (18 September 1995). Logging the globe. McGill-Queen's Press – MQUP. pp. 157–. ISBN 978-0-7735-1346-4. Retrieved 4 December 2011.
178. Biodiversity and Infectious Diseases | The Center for Health and the Global Environment. Chgeharvard.org (2011-11-11). Retrieved 2016-11-13.
179. Forests and human health. Fao.org. Retrieved 2016-11-13.
180. Moslemi, Jennifer M.; Snider, Sunny B.; MacNeill, Keeley; Gilliam, James F.; Flecker, Alexander S. (2012). "Impacts of an Invasive Snail". PLoS ONE. 7 (6): e38806. doi:10.1371/journal.pone.0038806. PMC 3382606 . PMID 22761706.

181. Deforestation and emerging diseases | Bulletin of the Atomic Scientists. Thebulletin.org (2011-02-15). Retrieved 2016-11-13.

182. African Politics Portal | Tag Archive | Environmental impact of deforestation in Kenya. African-politics.com (2009-05-28). Retrieved 2016-11-13.

183. 2014 Kenya Economic Survey Marks Malaria As Country's Leading Cause Of Death | The Henry J. Kaiser Family Foundation. Kff.org (2014-05-01). Retrieved 2016-11-13.

184. Scheidt, Spencer N.; Hurlbert, Allen H. (2014). "Range Expansion and Population Dynamics of an Invasive Species: The Eurasian Collared-Dove (Streptopelia decaocto)". PLoS ONE. 9 (10): e111510. doi:10.1371/journal.pone.0111510. PMC 4213033 . PMID 25354270.

185. Lam, Sai Kit; Chua, Kaw Bing (2002). "Nipah Virus Encephalitis Outbreak in Malaysia". Clinical Infectious Diseases. 34: S48–51. doi:10.1086/338818. PMID 11938496.

186. "Nipah Virus (NiV)". cdc.gov.

187. Deforestation sparks giant rodent invasions. News.mongabay.com (2010-12-15). Retrieved 2016-11-13.

Index

R

S

T

W